Our Universe

The Universe in which we live is unimaginably vast and ancient, with countless star-systems, galaxies and extraordinary phenomena as strange and different from each other as black holes, dark-matter, quasars, gamma-ray bursts and diffuse nearly-invisible galaxies. From our earliest days humankind has looked to the heavens and wondered in awe.

Our Universe is a fascinating collection of essays on extragalactic astronomy and cosmology at the dawn of the twenty-first century. This is the second in a series of extraordinary books in which S. Alan Stern has brought together leading space scientists to describe their work. The first of these, *Our Worlds*, looked at the faraway worlds of our solar system, but in this new book we leave our sun behind to explore the vastness of the Universe itself.

This accessible and wonderfully illustrated book has been written by some of the world's foremost astrophysicists. Some are theorists, some computational modellers, some observers, but all provide deep insight into the most cutting-edge, difficult, and bizarre topics of all astrophysics.

Once again, highly personal accounts reveal much more than the wonders and achievements of modern astronomy, more than just its techniques and state of knowledge. *Our Universe* also gives unique perspectives on what drives these extraordinary, talented scientists and how their careers and very lives have been shaped by a burning desire to understand our Universe.

To my parents and their generation,
for opening the door

Our Universe

The Thrill of
Extragalactic Exploration
As Told by Leading Experts

Edited by
S. ALAN STERN

CAMBRIDGE
UNIVERSITY PRESS

PUBLISHED BY THE PRESS SYNDICATE OF THE UNIVERSITY OF CAMBRIDGE
The Pitt Building, Trumpington Street, Cambridge, United Kingdom

CAMBRIDGE UNIVERSITY PRESS
The Edinburgh Building, Cambridge CB2 2RU, UK
40 West 20th Street, New York, NY 10011-4211, USA
10 Stamford Road, Oakleigh, VIC 3166, Australia
Ruiz de Alarcón 13, 28014 Madrid, Spain
Dock House, The Waterfront, Cape Town 8001, South Africa

http://www.cambridge.org

First published 2001

Printed in the United Kingdom at the University Press, Cambridge

Typeset in Hollander regular 10/15pt [VN]

A catalogue record for this book is available from the British Library

ISBN 0 521 78330 5 hardback
ISBN 0 521 78907 9 paperback

Contents

Contents

Preface

Modern human civilization now stretches back almost 300 generations to the earliest organized cities. For most of that time, each clutch of humans identified their settlement and its surrounds as their home. Less than 100 generations ago, information transmission and transportation technologies were capable enough for people to form nation-states consisting of many cities and villages and consider them as a new kind of "home." In the last two generations—with the advent of space travel—many people have come to see their "home" as the whole of the Earth. This is an idea that would have been unthinkable to the ancients—for the world was too large for their technology to integrate the world, or even a nation-state, into an accessible and cohesive community.

So too, though it may not be hard in the future, it is hard for us, now, to think of our "home" as being something larger than our planet. After all, we are still trapped, both physically and to a very great degree intellectually, on our wonderful home, this planet, Earth. A century ago, Konstantine Tsiolkovsky, the great Russian space visionary, described the Earth as the cradle of mankind, saying that humankind, like any infant, cannot live in its cradle forever.

For perhaps at best a few thousand humans (about one in every 10 million), those who are planetary scientists, astrophysicists, extragalactic astronomers, and cosmologists, this vision of the Earth as a cradle from which we peer outward to learn about a larger realm is already becoming a familiar and natural concept. And it is this concept, in part, that gave birth to the idea in 1996 to bring together a few of the very best planetary scientists in the world to write about their favorite worlds, and in doing so to give a little perspective on what makes both them, and their favorite places, tick. That effort culminated in the publication of a book of essays about planetary science by planetary scientists. The book, called *Our Worlds*, was published by Cambridge University Press in 1998. On the heels of that volume, I wanted also to tell some stories of extragalactic and cosmological exploration though the eyes of the scientists who are charting that vast and deep ocean of space and time.

And so, with the blessing of Simon Mitton, the director of astronomical publications at Cambridge University Press, I set out in late 1998 to contact some well-known stars among the firmament of astronomical researchers, and asked each to tell a personal story involving their own career and motivations, and to describe some part of a favorite topic in which they had invested long years exploring. I asked each to tell their story from the heart.

What follows in this volume is a set of nine wonderful and diverse essays ranging across the breadth of extragalactic astronomy. The stories in this book, *Our Universe*, range from giant lurking galaxies so diffuse they were hardly known until recent years, to the fireball of the Big Bang, to the hearts of black holes. Within this book you will find both a lot of modern astrophysical science, and an insider's perspective about how turn-of-the-century astrophysics is done. You will also see a good deal of what drives and interests lifelong astronomers and, on occasion, you will learn something about their inner hopes and aspirations.

So, come and listen to some personal tales of human exploration, high-tech gadgetry, and the thrill of being a detective to the Universe. Come and visit *Our Universe*.

Alan Stern
November, 1999

The frontier Universe: At the edge of the night

ALAN STERN, Southwest Research Institute

The place we call our Universe is, for the most part, cold and dark and all but endless. It is the emptiest of empties. It is old, and yet young. It contains much that is dead, and yet much that is alive, forever re-inventing itself, and sometimes inventing something wholly new. It is permeated in a vacuum more than a billion, billion times the rarest air that ever wafts across the peak of Earth's Everest. This vacuum, though an insulator to sound and material communication (in part owing to its unimaginable expanse), is clearer by many orders of magnitude than the clearest Colorado sky, and so transmits across itself rich signals of light and gravity that reveal both its present-day and ancient workings.

Our Universe is larger than we humans can comprehend in any real sense, and it contains all we know. It is to our time what the Earth alone was to the ancients, but it is more, as well. For this magical place, this *all*, this *Universe* is also a source of inspiration, awe, and wonder that few humans can resist when they truly contemplate its meaning. So too it is the home of our home world, our home Solar System, our home Galaxy, and very likely all that we as a species and a civilization shall ever comprehend.

Within the depths of our Universe lie countless galaxies, and within each galaxy countless star systems and even more countless planets. Our Universe contains radiations that, while bathing it in light, poison many of its locales against anything we mortal shells of carbon and water could ever survive. And beyond the light and radiation, our Universe very likely contains inhabitants as bizarre and different from one another as black holes, quasars, roses, and the lurking hulks of faint, diffuse galaxies as ghostly as any Transilvanian fog. Perhaps the Universe also contains countless examples of life that are self-organizing and sometimes self-aware, counter to Nature's entropy. Perhaps not. We do not yet know.

In a scientific sense, we humans have only known that there even *is* a universe in the space beyond the Earth for a few handfuls of generations—far less than 1% of the time our species has walked this green Earth. And we, alone of Earth's creatures, perhaps (shudder the thought!) we alone of all creatures here and everywhere, have looked up beyond the sky, into the arms of this Universe, and asked the reporter's questions: "what and how, and when, and why?"

Those of us who ponder these questions, The Astronomers, wish to shed light on no less than *everything* in creation. It is a task so audacious that few of us, myself included, could sleep at night if we contemplated the challenge completely. (What ant in Manhattan wonders about, or is even aware of, the city that surrounds it?) We, of the cinder of stars, the debris of Solar System formation, and later of "slime mold" that rose a hundred thousand flights of biological stairs to become mammalia, and then further, *homo sapiens*, we dare to ask whence we came and what the Universe consists of? How does the "inside of the clock" work?

Nevertheless, we astronomers, practitioners of one of Earth's oldest professions, do dare to ask. And we do with increasing voracity seem to be progressing toward a real, if approximate, understanding of our Universe. It is more than any other species has achieved regarding its place on Earth, much less Earth's place in the all that the Universe is. It is something that sets humankind apart from whence we came.

In the collection of essays that follows, nine noted and accomplished extragalactic astronomers and cosmologists—some specialists in theory, some computational modelers, some observers—have written essays about their dearest subjects of study. In doing so, they have set forth watermark explanations of the state of knowledge regarding some of the hottest, most difficult, and most bizarre topics in all of far-away astrophysics. And so too, these nine experts have written about *why* their topics are of interest to them, why their careers, and even their lives, have been shaped by their particular cosmic quests.

In reading this collection of essays you will learn a good deal of the inner workings of modern astronomy, and its techniques and its state of knowledge. But you will also learn a good deal about the inner workings of a few of its most noted practitioners.

So come and follow along now, as a very special and talented nine people explain the latest explorations of no less than a Universe.

PART I
Revealing a Universe

1

Mapmaker, mapmaker make me a map

Probing the Universe's large-scale structure

JOHN HUCHRA, Harvard-Smithsonian Center for Astrophysics

John Huchra is one of the most naturally gifted extragalactic observers working today. He was educated in physics at MIT (Massachusetts Institute of Technology) and earned his PhD from Caltech (California Institute of Technology), but has spent most of his professional career at Harvard-Smithsonian. John's interests span cosmology, galaxy cluster dynamics, the large structure in the Universe, and star formation across the Universe. John is an avid outdoorsman, enjoying hiking, canoeing, and skiing. He and his wife Rebecca Henderson live in Lexington, Massachusetts, with their young son, Harry. John's specialty is doing large-scale projects in a field more often dominated by one- and two-person teams, something he tells us about here.

I love being on mountaintops. It's the next best thing to being in space. I guess I also love counting things, whether the things are 4,000 footers in New England, cards in games of chance, or galaxies on my observing list. Therein, of course, lies the tale.

It all started because I was a little kid much more interested in reading than in sports. I grew up in a moderately rough, poor neighborhood in northern New Jersey just outside New York City. I was lucky that both my parents were quite intelligent and always stressed the value of hard work and knowledge. That got me into reading, and science and science fiction were at the top of my list. By eleven I was trying to decipher *One, Two, Three Infinity* and *The Birth and Death of the Sun* by George Gamow and Fred Hoyle respectively. These books were amusing and described a beautiful and mind-stretching subject, so I knew quite early on that physics or mathematics was what I wanted to do. I took every opportunity I could to learn and experience more about science. In high school one summer I went to a camp to study ecology and conservation. The next summer I went to a wonderful National Science Foundation (NSF) summer program on chemistry at what was then the Newark College of Engineering. I learned how to program, studied the "vapor pressure of organic borate esters," and baked brownies in the ovens in the chemistry laboratory. I even took Latin and drafting, figuring that a scientist ought to be able to name things and draw her or his experiments.

MIT came next. I had a slow start—it's interesting to walk into an auditorium and hear the famous lecture that starts "look to your right, look to your left, one of you three won't be here in four years time," have the entering classes SATs (Scholastic Aptitude Tests) posted and realize you were in the bottom third of the class. Fortunately I managed to ignore that and keep on with the dream of being a scientist. I took a wonderful freshman seminar on cosmology with Philip Morrison (a course I now sometimes teach myself). Included in the usual under-graduate requisites was spending time as the social chairman for a dormitory of 550 guys. I eventually ended up playing with both modula-tion collimators for X-ray astronomy rocket experiments and stellar pulsation codes, that last as an undergraduate thesis topic with Icko Iben, now of the University of Illinois. I'm still not sure exactly why, but I found myself drifting slowly away from theoretical physics and into astronomy and astrophysics.

A few strokes of luck further firmed that career path. The first was flunking my draft physical—although that was a mixed blessing that I had to pay for a decade later with a cornea transplant—and the second was getting into Caltech. I went with the expectation of becoming a theorist, but that was just not to be. At every crossroad, I found myself moving more and more to the observational side. First, my fellowship paid $200 per month, and the rent was $125 per month. To solve that problem, I took a research assistantship helping to build a pressured scanned Fabry-Perot spectrometer for studies of planetary atmos-pheres. I chose a research project with George Preston on the measure-ment of magnetic fields in "peculiar A stars." When that project was completed and the time came to pick a thesis advisor and topic, Wal Sargent took me under his wing. I started working on galaxies, little blue ones to be exact.

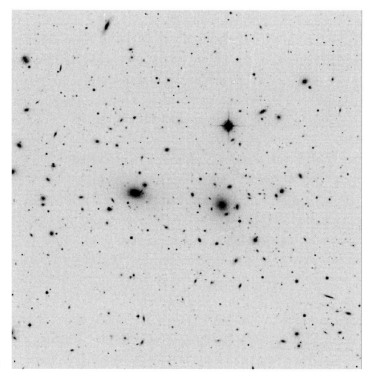

Figure 1.1

The Coma cluster of galaxies. (Image from the digitized Palomar Sky Survey, produced by the Space Telescope Science Institute, AURA Inc.)

Wal also offered me another great opportunity, to work on the Palomar Supernovae Search, which had been started by the noted Caltech astrophysicist Fritz Zwicky decades earlier. Although this was generally a hard and somewhat boring task, involving observing (with the 18-inch and sometimes the 48-inch diameter Schmidt telescopes) at Palomar for 5–7 nights a month, anyone doing it successfully quickly learned a lot about telescopes, photographic plates, the sky in general, galaxies and galaxy clusters in particular, and, most importantly, about patience. I enjoyed the solitude and the occasional thrill of discovery. I got into making improvements in the observing system. I was really enjoying doing something I was good at.

I was home. Between the supernova (SN) search and observing for my thesis project, I was observing ten or more nights a month. I found other excuses to go observing as well. I became the student in charge of checking out new visiting astronomers on all the small telescopes (the

200-inch was still the exclusive province of the senior staff, and students generally did not do independent work that required 200-inch time) and, whenever one of my fellow students needed an assistant, I was a willing volunteer. Observing at Palomar and Mt. Wilson was also a great way to meet people. Famous astronomers (and many who would later become famous) came as visitors, and the senior staff at Carnegie, people like Allan Sandage, Leonard Searle, Olin Wilson and George Preston, who rarely came down to Caltech, were lunch and dinner mates at the "Monasteries" (so-called because, up until the 1970s, women were not allowed to stay at them). It was still the case that the 200-inch or 100-inch observers sat at the heads of their respective tables, with fellow observers arrayed down the sides in order of telescope size. That generally kept us students in our place but, every once in a while, I got to be the 100-inch observer when a senior professor was on the 60-inch. Conversations were heady with the latest discoveries as well as the usual gossip and politics.

Through the supernovae search observations, I discovered a comet (1973h) and recovered one of the Mars-orbit-crossing asteroids that had been discovered three and a half decades earlier but then lost. Even that led to a job helping the geologists use the Schmidt telescopes for asteroid searches. Gene Shoemaker and Elinor Helin had developed a new and intense interest in finding Earth-orbit-crossing objects to bolster their theory that cratering on the Moon (and Earth!) was of asteroidal rather than volcanic origin. They needed to learn to use the Schmidt telescopes, and I was the man.

It's hard to describe the beauty of observing in those days. I was lucky to be able to make use of the Hooker 100-inch telescope at Mt. Wilson for my thesis. Completed in 1917, the 100-inch had a glass mirror that had been superbly figured by G. Ritchey. You would observe galaxies from the optically fast Newtonian focus at the top of the telescope, standing on a platform 40 to 50 feet above the floor of the dome. In the late summer and fall, the city of Los Angeles would blaze in glory (ugh!) outside the dome. But, in the spring, Mt. Wilson could be one of the darkest sites I've ever observed at; the Pacific fog would cover the city so that you could often see the tops of the clouds illuminated from above by starlight. I was not only home, I was hooked.

On being an astronomer

Perhaps the hardest transition in science is moving from the life of a graduate student to that of an independent scientist. The object of the game is to go from working on *one* problem that has primarily been posed for you to being able to pose exciting and tractable problems yourself. One thing we all must learn if we are to succeed is that ideas are the coin of the realm. And really good ideas are not easy to come by. The following story illustrates the point.

Once upon a time in graduate school we had an astronomy department retreat for the faculty, postdocs, and students. It rained. Almost by definition, we ended up in a deep philosophical discussion concerning careers, and what made a successful scientist. We decided in the end that an individual's success in the game could be predicted by their characteristics in a seven-vector space. Each vector measured a critical personal characteristic or set of characteristics such as intelligence, taste and luck, and the ability to tell one's story (public relations). The vectors and their "unit" vectors, the people against which one was measured in astronomy in those days, were:

Raw Intelligence   S. Chandrasekhar
Knowledge   A. Sandage
Public Relations   C. Sagan
Creativity   J. Ostriker
Taste   W. Sargent
Effectiveness   J. Gunn
Competence   M. Schmidt

(Here, I've changed a few names to protect the innocent.)

Each unit vector represented someone who was without equal at the time (1974 or so), for example Chandrasekhar was the smartest person in astronomy any of us had come across, and similarly, Allan Sandage represented the unit vector of knowledge (which is *not* the same as intelligence, although he is a damn smart cookie!). Some vectors are worth more than others, for example taste and creativity are probably more important than knowledge. Looking back on this I've come to realize that being nearly a unit vector in any one of the important characteristics almost guarantees you a tenured job, two are good for membership in the National Academy, and three put you in contention for the Nobel Prize.[1]

So how do you go about developing some of these qualities? Again I was aided by a few flukes of fate. One was that I'd accepted a job in Australia, but, before I finished defending my thesis, there was a vote of no confidence for Australia's Prime Minister of that time and the government had to be reorganized. That froze all government jobs including mine. So, with my thesis 99% completed, I had no job.

Rather than turning my thesis in at that point, I stopped to think about the problem, something I hadn't been able to do while madly collecting data and writing it up. I also had the time to think about other, new projects, some related to my research, and some very interesting sidelines.

What to do after graduate school

My second fluke was falling into one of these new projects. At that time, a number of people at Caltech had come from Princeton where they'd been influenced by Jim Peebles, one of the great theoretical cosmologists of our time. Peebles had started trying to understand the clustering of galaxies, basically how the Universe got from a pretty smooth and uniform state at the time of the formation of the cosmic microwave background to the details of galaxy clustering we see today. His target was not just galaxy formation, a hard enough problem in itself, but the formation of large-scale structure. However, in trying to think about this, he had a simple problem, one even such as I could try to deal with. In the early 1970s many galaxy catalogs existed, based primarily on identifying galaxies by eye from the large photographic sky surveys of the 1950s and 1960s. Two examples were the Shane–Wirtanen catalog, which is not a set of actual galaxy identifications but a fine grid of galaxy counts of about 1,000,000 objects made from astrographic plates

---

[1] Many people would want to add "luck" to the list, but our learned conclusion was that luck is a product of at least three of the above vectors and not an attribute in and of itself.

from Lick Observatory, and the infamous Zwicky catalog (where "infamous" applies both to Zwicky and his catalog).[2]

The Zwicky catalog is a list of more than 31,000 objects identified on the 14×14 inch glass plates of the Palomar Schmidt telescope. Peebles thus had lots of grist to study the galaxy distribution in two dimensions, but the structures were sure to be three dimensional and there was almost nothing known about the 3-D galaxy distribution. There were hints of filamentary structures in the maps Peebles and his coworkers made from both those catalogs, but what is filamentary on the sky could look quite different in 3-D. Projection effects also quite easily wash out the finer details of 3-D structures.

Although it had been known from Edwin Hubble's work in the late 1920s that there is a very good linear relation between the apparent radial velocity or redshift of a galaxy and its distance away from us (another long story for a different book!), measuring redshifts for large numbers of faint galaxies was quite a chore in the early 1970s. In fact, in 1972, after more than 60 years of observing galaxy spectra by the likes of Vesto Slipher, Milton Humason, and Gerard deVaucouleurs, there were only a little over 1,000 galaxy redshifts known, and many of these were for galaxies in clusters of galaxies. The largest complete, brightness-limited redshift catalog contained less than 300 galaxies. Even though the method for creating a 3-D map of the galaxy distribution existed, the tools to apply this method were too primitive. Astronomers were just beginning to move away from photographic plates, which actually detect less than 1% of the light that falls on them, to much more efficient electronic detectors. Time on big telescopes was hard to come

---

[2] Fritz Zwicky was a professor of astrophysics at Caltech and one of the true giants of twentieth-century astronomy. He was responsible not only for producing one of the most important catalogs of galaxies, with over 31,000 entries, but also, among many other things, for predicting the existence of neutron stars and gravitational lensing, for recognizing the importance of clustering in the Universe, especially galaxy clusters themselves, compact galaxies, wide field imaging, and supernovae, including organizing the first major supernova search at Palomar Observatory. He was the discoverer of "dark matter." He also served as a major foil for first Edwin Hubble, and later Hubble's protegé, Allan Sandage. The introduction to his "Red Volume" (*A Catalogue of Compact and Post Eruptive Galaxies*) is essential reading for any student not destined to be one of the "high priests" or "sycophants" of American astronomy.

by (there were only the Hale 200-inch telescope at Palomar, the 120-inch at Lick, and the 107-inch at McDonald—the 4-meter telescopes at the national observatories hadn't been built yet), and small telescopes were generally not equipped with "modern" detectors. Nonetheless, Peebles began urging people to think about redshift surveys as a way of making real 3-D maps.

At that time (as today) the Princeton–Caltech axis was strong, with lots of trading of students and postdocs (postdoctoral researchers) back and forth. J. Richard Gott, who had been a student at Princeton, came to work at Caltech as a postdoc. He and Ed Turner, a fellow student, took to analyzing the clustering properties of the galaxies in Zwicky's catalog using whatever tools they could, including the small number of redshifts available. They produced a beautiful series of papers analyzing the properties of a complete, brightness-limited sample of galaxies but with one flaw—the analysis was perforce only two dimensional. They still didn't have the tools to make a 3-D map.

Redshift surveys

The break in the dam occurred in the mid-1970s. First, electronic detectors (in the form of image intensifier tubes) became available commercially and cheaply enough for small telescopes to be equipped with them. Image tubes have 20 times the efficiency of the best astronomical photographic plate, and made a 1-meter diameter telescope the equivalent of the 200-inch! Second, centimeter wave radio receivers increased in sensitivity and the first efficient radio spectral correlators came into operation. This enabled rapid and accurate determination of the redshifts of galaxies with lots of neutral hydrogen gas, the most common element in the Universe and a major constituent of every spiral and irregular galaxy. In the space of just a few years, a redshift survey of 1,000 or more galaxies went from being a daunting and near-impossible task to merely a very difficult one.

As both Ed and Richard prepared to go to Princeton, Ed to the Institute for Advanced Study, and Richard back to Princeton University, another collaboration was formed to attack this problem by obtaining redshifts for the Turner–Gott sample. The leaders were Wal Sargent, Gill

Knapp and Trinh Thuan. Fortunately for me, they needed a little more horsepower, and I had both a little extra time on my hands and also a reasonably high "competence" score, that is I could make telescopes sit up and take data. We started in the winter of 1975 to get redshifts for the spiral galaxies at Green Bank and Arecibo, two of the modern radio facilities, and at Palomar and later Kitt Peak for the gas-poor elliptical and lenticular galaxies. Our goal was to obtain redshifts (nee distances) and accurate brightness measurements for 1,100 or so galaxies. We made a fairly good start of it, too, obtaining about 650 new redshifts in a year and a half before that collaboration broke up. All the players found themselves at different, widely scattered places, and e-mail hadn't been invented yet!

Meanwhile, I was aided by another bit of luck. On the job market once again, but this time with more publications and actually a much better idea of what I wanted to do next, I landed a postdoctoral position at the newly formed Center for Astrophysics in Cambridge, Massachusetts. That was surprising because I'd hoped and expected to get a job at Kitt Peak National Observatory (KPNO), the home to many a young observer, but that year KPNO chose a theorist instead. The Harvard-Smithsonian Center for Astronomy, or CfA, which had a reputation for hiring theorists, that year hired two observers! At CfA, I met Marc Davis and Margaret Geller, two other Princeton-trained theorists who had worked or were working with Jim Peebles. Marc was desperately trying to get his own redshift survey started, but he was having limited success. The existing instrument at CfA's observatory on Mt. Hopkins in Arizona had some real problems. Marc wanted to buy a very expensive camera from a commercial firm, but there was little money coming from either the CfA or the NSF to do so. Marc and Margaret and I started working on the best data set I could assemble, about 1,200 redshifts for fairly bright galaxies (see Figure 1.2a), but the only progress being made on new observations was the observing at the Kitt Peak 0.9-m telescope and the Green Bank 300-ft telescope for the Caltech consortium, soon to wind down.

Then came yet another statistical accident. I was at Palomar on a cloudy night, strolling the catwalk of the 200-inch telescope in the fog with Steve Shectman and talking about spectra and redshifts. Steve, one

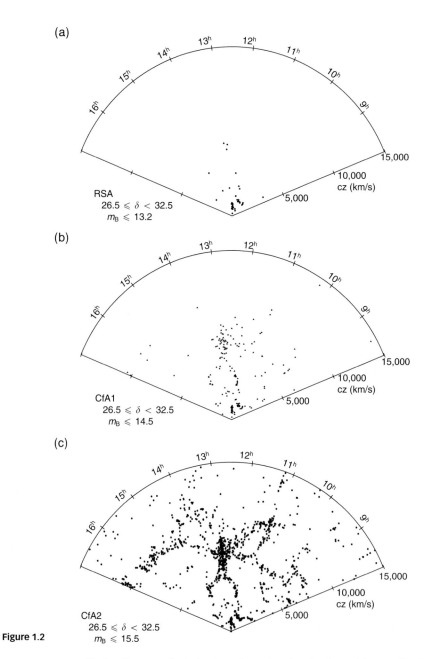

Figure 1.2

Wedge diagrams displaying two-dimensional projections of the 3-D distribution of galaxies. Each plot shows the distribution of galaxies in a wedge of sky six degrees thick by ~100 degrees wide. (a) The view in 1978, before the start of the first CfA Survey. (b) The view in 1982, after the completion of the first CfA Survey by Davis, Huchra, Latham and Tonry. (c) The view in 1985, after the completion of the first CfA2 Survey strip by de Lapparent, Geller and Huchra.

of the real unsung heroes of redshift survey work, described a new instrument he was building, a photon-counting Reticon® detector for spectroscopy, which sounded a lot more robust than the commercial system and was one tenth of the cost. Steve offered to lend us the plans and even let us go to his lab and copy the instrument electronics. This sounded like a match made in heaven. I went back to Cambridge, convinced Marc (it didn't take much!) that this was the way to go, and he and I proceeded to persuade our bosses to make some resources available to make it happen. Enter other heros: Herb Gursky, then the Associate Director at CfA, who came up with the internal support, including a few good computer programmers, and George Field, the CfA Director, who went to bat at the NSF to get Marc additional funds. Thus, in the early summer of 1977 the first CfA Redshift Survey was born.

Marc was in charge of duplicating the camera and electronics, and Tom Stevenson, the lead computer programmer, was in charge of data taking and analysis system. John Tonry (Marc's graduate student) wrote the data analysis software for our Data General Nova computers (then state of the art). Dave Latham, a Smithsonian Astrophysical Observatory (SAO) scientist, took the lead in putting together the front of the detector package, a three-stage image-intensifier tube. Dave and I, with the Mt. Hopkins staff, worked over the existing spectrograph, and found and corrected numerous optical problems including a misplaced collimator and mirrors that vignetted the field of view. I assembled the catalog of objects to be observed, all the galaxies brighter than a blue magnitude of 14.5 in Zwicky's catalog of galaxies and at high galactic latitude. Our own galaxy, the Milky Way, is a spiral galaxy that contains lots of gas and dust in its disk. The Solar System sits almost exactly in the plane of the Milky Way. As a consequence, when we astronomers look out at the Universe, we get by far the clearest view when we look perpendicular to the plane of the Galaxy, at high galactic latitude.

We took our first real data in the spring of 1979 using the mighty(!) 24-inch telescope at Mt. Hopkins. We quite quickly switched to the 1.5-meter, and, because the 1.5-m was a really terrible telescope with a spherical mirror, we got to use it for spectroscopy a very large chunk of time. The Gary Larson cartoon showing a bunch of astronomers monopolizing the telescope all day (sic) long was not far from the truth. When the Moon was down, one of us, often me, was at the telescope. At the same time, I was also heavily involved with Marc Aaronson and Jeremy Mould on a new project to measure the Hubble constant with the infrared Tully–Fisher relation, a hot topic even today, for which we got lots of time at the small telescopes at the National Observatories. I often spent six weeks at a time on mountaintops around Tucson, first measuring redshifts, then doing infrared photometry of galaxies and then going back to redshifts as the Moon waxed and waned. In retrospect, it was also probably the only way to survive on a postdoctoral or junior staff salary—canned food in the mountain microwave usually covered by a too-small expense allowance. But it did improve my cooking skills!

I took the last redshift for that first survey of 2,400 sources (2,399 galaxies and 1 star) in June of 1981. We performed a number of statistical analyses of the data, getting perhaps the best galaxy luminosity function (the number of galaxies of a given luminosity per unit volume) and galaxy correlation function of the time. We measured the overdensity of the local supercluster of galaxies and saw very strong hints of larger-scale structure, but even a survey of 2,400 galaxies wasn't very deep. Figure 1.2b shows the state of the map at the end of the first CfA survey for the volume shown in Figure 1.2a. Despite a significant improvement, the maps weren't really crisp and clear. As a result, none of the results around, including those from other surveys, really could convince either the theorists or observers that much was happening beyond Hubble's "sensibly uniform" universe.

The second CfA Survey

About then, Marc Davis went off to Berkeley and Margaret Geller got back to CfA from an extended stay at the University of Cambridge. Margaret and I started working on what is called the Omega problem (determining the mean mass density of the Universe to see if it's going to collapse back on itself or expand forever) by selecting groups of galaxies from the redshift survey and by measuring masses for groups and clusters of galaxies. We made some interesting discoveries about the dynamical state and relative youth of nearby galaxy clusters—we are living in the time of cluster formation—driven by Margaret's studies of cluster substructure and my work on the Virgo cluster. But in a few years we began to reach the end of what could profitably be observed in galaxy clusters with the 1.5-m telescope. Both Margaret and I had been dreaming of a wide-field redshift survey much deeper than the first CfA Survey, but both of us were wary of asking for so much telescope time.

Finally, in 1985, we took the plunge. The 1.5-m was an excellent spectroscopic telescope, it could easily reach a magnitude fainter than the first CfA Survey, and Zwicky's catalog quite nicely went to that depth. Instead of a paltry 2,400 galaxies, we decided to go after 15,000. The problem was that there was a great debate about the methodology of the next survey. There were essentially three plans floating around. Marc Davis suggested a knitting needle approach, namely sampling one-in-five or one-in-ten of the fainter galaxies to increase the volume surveyed very rapidly, but not so densely. Simon White, another player in the game, wanted dense sampling but in a smaller, contiguous, square or rectangular area of the sky. Margaret was convinced that long and relatively thin strips across the sky were the way to go.

You can think of the mapping problem this way. Suppose you want to study the topography of the surface of the Earth, and you have a steerable satellite but only a limited amount of film, say enough to take pictures of 1,000 square miles. You could take random 1 square mile shots of the surface (the Marc Davis approach), you could carefully image a 33×33 mile square (the Simon White approach) or you could try to observe a strip, say 5,000 miles by one fifth of a mile (the Margaret Geller approach). The first approach would give you a fairly good idea of the fraction of the Earth's surface covered by ocean, desert, mountains, etc., but you wouldn't know anything about the sizes of such things. This type of sparse sampling was actually used for one of the earlier IRAS (Infrared Astronomy Satellite) galaxy redshift surveys, the QDOT (Queen Mary/Durham/Oxford & Toronto) survey, and produces a deep but very low resolution map, fine for continents, but watch out for mountain ranges! The second approach would give you very detailed information about a specific place, but since you're likely to see only ocean or desert or mountain, you'd have a very distorted view of the

Earth. The third approach, however, is a winner, since not only are you likely to cross a little bit of everything, but you can also estimate the sizes of the oceans, continents and mountain ranges you cross. Not a map, but surely a mapmaker's first crude topographical survey!

Better yet, from the point of view of a practiced observer, since the Earth rotates and the sky swings around overhead, if you observe in strips you can use the telescope in the most efficient manner—fewer slews, less time calibrating, all those good things. To me, that, plus Margaret's argument on sampling structure sizes, made the case.

I started observing for the second CfA Redshift Survey in the winter of 1984/85. Valerie deLapparent, a graduate student working with us on large-scale structure and galaxy surveys, made some of the observations and also was given the task of plotting up the data from our first strip. I was so sure that we wouldn't find anything extraordinary (and also too tied up taking the data and too lazy to plot it as we went along!) that the plots of the galaxy distribution weren't made until all the data were in hand. When Valerie showed me the initial map in June of 1985, my first, very conservative reaction was "whoops! what did I do wrong?". But repeated checks showed that our map, Figure 1.2c, was right. With this much deeper and denser map, we saw that the galaxy distribution was far from random. It was not the "meat ball" topology of lumps (clusters) in an otherwise uniform soup (the field) that was the favorite view of theorists at the time. Galaxies lay on surfaces surrounding large empty regions, "voids." The first large void had actually been discovered four years earlier by Kirshner, Oemler, Schechter and Shectman, but had been dismissed as a fluke, a "one-off," that would not be seen again. But in one season's observing on the "mighty" 1.5-m telescope, we shot that idea to pieces! The Universe was frothy. Most of the Universe was filled with voids. Margaret coined the analogy of the soap bubble universe, and it stuck.

The CfA2 Redshift Survey is now complete. We have measured redshifts for over 18,000 bright galaxies in the northern celestial hemisphere. We have exquisite views of the nearby galaxy distribution over small volumes of space (Figures 3 and 4, color section). And we know that the distribution of galaxies is exceedingly non-random. Galaxies are distributed on quilt-like surfaces, with lumps that are the large galaxy clusters at intersections of these surfaces. With 24 slices in hand, the basic results from our first slice still hold. We discovered the Great Wall of galaxies in 1989, and it remains one of the largest structures ever seen in the Universe. Theorists, using N-body gravitational simulations (see Nick Gnedin's chapter), are still trying to match the observed galaxy distribution with all the physics they can muster and with arcane mixes of normal baryons (the stuff you and I are made of), cold dark matter, neutrinos and even a Cosmological Constant, but they haven't been fully successful yet and there's still a lot of work to do.[3]

The next maps—more to come

By definition, observationally we still have a long way to go. Existing redshift surveys have mapped only a very small percentage of the volume of the Universe and not very well at that. Most surveys are based on catalogs of galaxies assembled from photographic plates. These catalogs are non-uniform and not very accurate. Large chunks of the sky are invisible because they are obscured in visible light by the gas and dust of our own galaxy.

However, several new large surveys are underway. One, called the SDSS, Sloan Digital Sky Survey, aims to use a 2.5-m telescope in New Mexico to map about one fifth of the sky, both photometrically, with new digital imaging detectors, and spectroscopically, with fiber optically fed spectrographs. The plan is to measure 1,000,000 galaxy redshifts in the next five to ten years. The competition is the Two Degree Field Survey or 2DF, a survey of 250,000 galaxies, also over a small area of the sky, using a 3.9-m telescope in Australia, equipped with a special 2 degree field-of-view spectrograph (2DF). 2DF is underway and is madly trying to scoop SDSS. These surveys will provide excellent information about the statistics of galaxy clustering for matching to theories (and N-body simulations based on differing input physics) of structure and galaxy formation. However, they will still explore much less than one quarter of the sky.

Not being a real fan of maps with large areas marked "Here There Be

---

[3] More information on redshift surveys and maps of the local Universe can be found at John Huchra's website, http://cfa-www.harvard.edu/~huchra.

Dragons," my next project is an almost *whole sky* map of the *nearby* Universe. Again the game is to attack the Omega problem. How much does the Universe weigh?—what is its mean mass density? We know that our galaxy is moving at 630 km/s towards a point in the constellation of Hydra (see John Mather's chapter for more details on this). Can we identify the mass concentrations causing that motion? Can we match gravity, as measured by the motions of galaxies relative to the general expansion of the Universe, to local lumps of stuff—galaxies or otherwise—whose gravitational pull might cause such motions? In doing so, can we learn how much dark matter there is in the Universe and where it is located?

The project is called 2MASS, the Two Micron All Sky Survey, and it aims to use the penetrating power of infrared light to map the nearby distribution of galaxies at wavelengths unaffected by absorption by gas and dust in the Milky Way. My colleagues and I, led by Mike Skruskie and Ray Stiening at the University of Massachusetts, and including Steve Schneider, also at the University of Massachusettes, and Tom Jarrett, Tom Chester, Roc Cutri and Chas Beichman at Caltech's Infrared Processing and Analysis Center (IPAC), are using data from twin 1.3-m telescopes at Mt. Hopkins in Arizona and at Cerro Tololo in Chile. You need to observe with telescopes in both hemispheres to get the whole sky. As of mid-1999, we have over 1,400,000 galaxies in our catalog with only about 50% of the scans completed. We'll start by getting redshifts for the brightest 150,000 galaxies. Telescope time allocation committees willing, we'll eventually do 1,000,000 over the whole sky. This will be the deepest and highest resolution map of the whole local Universe ever made, providing reliable charts out to about 10% of the speed of light.

Its a voyage of discovery to the nearby Universe. My goal is to make the best map I can and leave no nearby areas uncovered. Come back in six to ten years and I'll show you the new geography!

2

Looking back in time: Searching for the most distant galaxies

ESTHER HU, University of Hawaii

Esther Hu was born and raised in New York City. She is a second generation Chinese-American whose parents came to the US as students at the end of the Second World War. Like her sister Evelyn, Esther decided to be a scientist before attending college. Esther was educated in physics at MIT and earned her PhD in astrophysics at Princeton. She then became a research associate with the X-ray group at NASA's Goddard Space Flight Center, and then a postdoctoral fellow at the Space Telescope Science Institute. She is now a professor of astronomy at the Institute for Astronomy at the University of Hawaii in Honolulu. In the course of her career, Esther has studied successively more distant objects across the Universe using more and more sensitive telescopes and instruments. Despite her friendly and easy-going nature, Esther is as competitive as they come; she presently holds the record for distant object detection. Esther enjoys reading, classical music, and "living in a place as beautiful as Hawaii."

The past is a foreign country: they do things differently there.

L.P. Hartley

When I was seven, at my first school book fair, I came away with a title, *Insight into Astronomy*. The "pull" behind the choice came from the quotation by Ralph Waldo Emerson in the preface: "If the stars should appear one night in a thousand years, how would men believe and adore, and preserve for many generations the remembrance . . ." The idea of vast cosmic distances, measured in light travel time, so that celestial objects are viewed through a kind of time machine, captured my imagination. What would an early snapshot of our own galaxy look like?

By the time I finished high school, the subject of astronomy had expanded to include studies of quasars, black holes, pulsars, relic radiation from the primordial fireball of the Big Bang, and other exotic phenomena, and now encompassed results from a growing space exploration program. Our observable Universe had become larger both in kind and extent. The most distant galaxies made up a frontier with a moving boundary—in more senses than one. Not only did the position of this boundary reflect distance limits continually being pushed back by new scientific observations, but the individual galaxies at these boundaries were themselves moving away from us.

The discovery that the recession speed of distant galaxies increases proportionately with their distance was made by the American astronomer, Edwin P. Hubble, in 1929, and is a result of the expansion of the Universe. Hubble used observed shifts in the frequency of light from galaxies to deduce their motions. Determining a source's motion by shifts in the frequency of its emitted light or sound is most familiar to us when we use the rising or falling pitch of an ambulance siren to judge whether the vehicle is approaching or moving away from us.

We perceive higher or lower frequency light as bluer or redder colors. Light from a receding galaxy is spoken of as "redshifted," and features, such as a pattern of emission or absorption lines, appear displaced to longer wavelengths. A galaxy's *redshift*, z, is defined as this increase in the wavelength of a feature expressed as a fraction of its wavelength when at rest. For nearby galaxies, the recession velocity is simply z multiplied by the speed of light.[1]

The expansion of the Universe causes recession velocity to increase with an object's distance, so the higher the redshift the farther away the galaxy. Distances to the highest redshift galaxies can also be translated through light travel times into "look-back times" in the early Universe.[2] The most distant galaxies discussed in Hubble's original paper had redshifts $z < 0.004$, or look-back times to when the Universe was 99.5% of its age—not very far back in time at all! Sixty years later, when a space

---

[1] This is approximately correct for redshifts much less than 1 ($z \ll 1$). For high-redshift (distant) galaxies, there is a relativistic correction, and the recession velocity, v, and redshift, z, are related by the equation:
$(1+z) = \sqrt{(c+v)/(c-v)}$, where c is the speed of light.

[2] The detailed transformation of redshift, z, into distance and look-back time also depends on how much the current expansion of the Universe has slowed. This deceleration term depends on the density of the Universe.

telescope named in Hubble's honor was launched, the highest measured galaxy redshifts were typically $z\sim1$—or a look-back to when the Universe was about a third of its present age. Viewed in terms of a person's lifetime, these redshifts show us galaxies as adults, not infants. The faintness of distant galaxies makes it difficult to determine their redshifts and to identify a high-redshift population without some way of making these objects stand out from the crowd. And the redshift–distance relation means that our view of distant galaxies is not only filtered through a time machine, it is also translated in wavelength.

The siren call of distant galaxies

Until very recently, the very high-redshift galaxies we knew about were unusual objects. Quasars and radio galaxies are a small fraction of the galaxy population. They're "screamers," and even in a babbling crowd you can pick them out at a distance by the volume and uniquely strained tones of their voices. The energy sources which power quasars can make them thousands of times brighter than normal galaxies in visible light, and consequently easier to study. These sources will also give them unusual colors, and can make them bright radio and X-ray sources as well. But we still needed a way to pick out the normal galaxies at high redshift.

In 1985, when I was in the second year of a postdoctoral fellowship at the newly constructed Space Telescope Science Institute (STScI) in Baltimore, George Djorgovski, Hy Spinrad, Pat McCarthy, and Michael Strauss of the University of California at Berkeley reported the discovery of a redshift $z = 3.215$ object near a bright quasar known to be at similar redshift—or a look-back to a Universe about 12% of its present age. An image of the quasar field had been taken through a narrow filter matched to the strongest quasar emission line: a signature of hydrogen dubbed "Lyman alpha." This line is expected to be the strongest feature for star-forming galaxies, as well as quasars, so the dramatic possibility was that we were seeing an early distant galaxy in the light of its forming stars. An even more exciting suggestion was that targeting known high-redshift objects was a good way of finding more high-redshift galaxies, since galaxies tend to cluster. And the additional emission-line galaxies discovered this way through a filter tuned to the redshifted quasar emission might be members of the normal high-redshift galaxy population. Studying such objects can tell us how typical galaxies form stars and evolve into galaxies like our own Milky Way Galaxy over the history of the Universe.

At the same time these scientific developments were taking place, I was having to make some career decisions about where I wanted to spend the next few years. Soon after I'd arrived at the Space Telescope Science Institute, the Deputy Director, Don Hall, left to become the Director of the Institute for Astronomy at the University of Hawaii. A number of Space Telescope postdoctoral researchers were to make this move but, at the time, moving to Hawaii with its superb ground-based telescope facilities had to be weighed against the excitement of being at STScI just after Hubble Space Telescope's (HST's) much-anticipated launch. Hawaii won, and by December 1985 I was making plans to arrive in Honolulu near the beginning of March 1986. About a month before I left Baltimore, the Challenger disaster struck, and one of the many consequences of this tragedy was that it would be another four years before HST had its first glimpse of the sky.[3]

To put the observational attractions of Hawaii in perspective, the problem of studying distant galaxies is that their images are small and faint, and you need to view these contrasted against the background night sky by making very deep observations (with very long exposures and/or a large telescope) and by minimizing atmospheric blurring of images. The summit of Mauna Kea in Hawaii routinely yields some of the best-quality astronomical images in the world, and regular access to the telescopes at this site made it feasible to undertake programs which were innovative but risky, or which required a long-term effort to bring off.

However, the first attempts to turn these resources to searches for

---

[3] On January 28, 1986 the Space Shuttle Challenger exploded upon launch, killing all seven crew members. At that time a substantial part of NASA's space program relied on space shuttles to deploy probes and observatories such as the Hubble Space Telescope. A highly readable account of the investigation into the Challenger accident is given in the book, *What Do You Care What Other People Think?* by Presidential Commission Member and Physics Nobel Laureate, Richard Feynman.

distant galaxies came up dry; none of the redshift $z > 3$ quasar fields used to target searches showed signs of companion high-redshift galaxies. Was this again because of the special nature of the original source? Yes, as it turned out; the original targeted quasar had been unusual in being a strong radio source as well, and radio galaxies have frequently been found with such surrounding emission. Selecting radio quasars as targets immediately turned up more of these quasar companions. All lay close enough to the quasar to be fueled by the unusual processes powering the radio source. None was likely to be an independent star-forming galaxy.

The technique of looking for very distant galaxies using the star-formation-powered, hydrogen ultraviolet emission line (called Lyman alpha), fell into disfavor. The cases where such emission had been detected were unusual, radio-loud objects and the general star-forming galaxy population had not been seen. In place of arguments that had touted Lyman alpha emission as a convincing way of finding high-redshift, star-forming galaxies, discussions now swung to explanations for the failure to detect emission from distant galaxies in the targeted quasar fields. The active star-forming regions of the Milky Way and nearby galaxies are located in parts of the spiral arms surrounded by dust and gas. It was argued that such dust, the product of generations of stellar processing of hydrogen into heavier elements, would easily block ultraviolet light; Lyman alpha emission in galaxies would be suppressed.

Taking different tacks—deep surveys with infrared detectors

While investigations of the quasar companions were going on, a new technological development—the birth of infrared imaging arrays for astronomy—started another line of investigation in the study of high-redshift galaxies. A difficulty in comparing present-day galaxies with distant ones is that emitted visible light gets redshifted to infrared wavelengths for high-redshift galaxies. Infrared cameras capable of surveying large areas of sky in the manner of optical cameras would become a potent force in distant galaxy studies. While earlier infrared detectors essentially consisted of single photocells, the new generation of detectors measured light with many such picture elements or "pixels" assembled in an array. The effect was like substituting an X-ray machine for a geiger counter to study a radioactive source: in place of a single reading of "counts" a detailed picture suddenly emerged of a source's shape and size, and of the variations in light intensity along its structure. With the initiation of Ian MacLean's (58×62 element) IRCAM infrared camera at the United Kingdom infrared 3.8-m telescope at the end of 1986, Len Cowie and Simon Lilly at the University of Hawaii decided to start deep imaging surveys of "blank sky" regions in multiple color bands. In much the same way that the "Hubble Deep Field" was to prove the starting point for a cosmic census some nine years later, these fields would be a way to study the evolution of galaxy populations. All optical color bands would be used, but the infrared data would be the unique and crucial part. Observations by successively larger and more sensitive infrared cameras used on a variety of Mauna Kea telescopes would sample distant galaxies where most of their light was being emitted.

The other key factor would be ongoing spectroscopic observations to identify the redshifts and types of galaxies in the Hawaii Survey Field samples. This was the most time-consuming part of the effort, and by the time these fields were observed with the Hubble Space Telescope, several hundred spectra and over six years of ground-based observations would have been spent in the preparation.

In late summer of 1991, a conversation at a workshop on high-redshift quasars hosted by the Space Telescope Science Institute would lead to the discovery of the first star-forming galaxies at redshifts $z > 4$ —or, roughly, within the first billion years in the life of the Universe. Richard McMahon, a Cambridge University astronomer, along with his colleagues had succeeded in identifying about twenty quasars at redshifts above 4. Together with the $z > 4$ quasars turned up by astronomers at Caltech over the preceding year and a half, these indicated there was likely to be a much larger population of ordinary faint galaxies present at these redshifts. We decided to look for these.

We try higher

Over the entrance to Hale Pohaku, the astronomers' midlevel office facility and dormitory on the Island of Hawaii, are the words, "We try higher." At over 13,700 feet (4,200 meters) above sea level, the summit of Mauna Kea is the highest site of any major observatory in the world.[4] The motto on the sign seemed appropriate for our project, which was to target a number of the highest-redshift quasars to look for distant, star-forming galaxies in the surrounding fields.

Galaxies tend to be found near other galaxies, and the idea was to use the quasars as signposts to mark the locations of their less spectacular neighbors. The most compelling reason for trying the emission-line search again was the much higher redshift of the quasar targets. Galaxies caught in the first outburst of star formation might not yet have formed and retained substantial amounts of dust, so the Lyman alpha emission line might be strong and visible. Secondly, although previous Lyman alpha emission-line searches of both "targeted" fields around quasars and "blank sky" regions had not been successful in identifying high-redshift galaxies, they had not used particularly deep exposures. Extrapolating the expected galaxy properties from what was known at redshift $z \sim 1$ to predict emission fluxes at redshifts $z = 3-4$ was a big stretch—the time interval was more than half the age of the Universe! The earlier failure to find emission at theoretically predicted values could reflect problems with these assumptions, rather than undetectable Lyman alpha fluxes. Using the University of Hawaii 2.2-m telescope, Richard McMahon and I took deep exposures targeted on the redshifted emission of half a dozen of the highest-redshift quasars.

These new searches turned up emission-line galaxies ten times fainter than the earlier quasar companions, and separated from the quasar by distances larger than our own Galaxy's separation from the neighboring Andromeda Galaxy. So these emission-line galaxies were too far away to be fueled by the unusual energy sources which powered the quasar, but were quite reasonably galaxies viewed in the light of their

[4] Construction of the Atacama Large Millimeter Array (ALMA) is scheduled to start in 2001. This international radio telescope facility will be located in Chile, at an altitude 16,400 feet (5,000 meters) above sea level. Oxygenated control rooms will be required for its operations, which could begin as early at 2005.

Figure 2.1 High-redshift galaxies at $z \sim 4.55$ (or about 10% of the age of the Universe) in a distant quasar field. The galaxies and the much brighter quasar are circled in the left panel, which is taken through a filter with a narrow bandpass centered on a strong ultraviolet hydrogen line, redshifted to red wavelengths. The right panel shows the same field through a broader filter which does not include the emission line, and in which the high-redshift galaxies are not visible. These images were taken on the University of Hawaii 2.2-m telescope with a 20-hour exposure through the narrow-band filter. Today, the left image could be taken with the Keck 10-m telescope in one hour. However, even with Keck, in a one-hour exposure away from the emission line these galaxies are barely detected.

own star-forming activity. However, the most convincing evidence that these were indeed high-redshift galaxies came from the high contrast which was observed between the brightness of emission viewed through the narrow-band filter compared with the galaxies' brightness at neighboring wavelengths (see Figure 2.1).

The crux of the argument is that in star-forming galaxies the energy funneled into an emission line and the light at wavelengths away from the line are both produced by the same source: stars. As a consequence, there's a natural maximum for the strength of an emission line relative to its surrounding continuum light, because only a fixed fraction of the energy in starlight can excite emission lines. However, the effect of redshift is not only to move the location of the line to longer (and redder) wavelengths, but also to increase its width by the same factor of $(1 + z)$. Highly redshifted emission features would now cover a broader fraction of the narrow-band filter (they would be said to have "high equivalent width"), while the featureless continuum away from the emission would look unchanged, and the observed line strength could be up to a factor of $(1+z)$ times higher than the natural maximum. The two important conclusions were: (a) the "high equivalent width" of the emission viewed through the narrow-band filter meant that we were dealing with very high-redshift galaxies if these were normal star-forming objects and (b) these emission-line strengths were so close to the maximum allowed values *even* with the $(1+z)$ multiplicative factors (around 5.5 for $z \sim 4.55$) that these objects could *not* be very dusty.

We had struck paydirt, and had found a way of turning up distant star-forming galaxies by using the emission in early outbursts of star formation to enhance their visibility! However, getting a closer look at these galaxies would be tough because these objects are so faint that they would be hard to study without a much larger telescope—such as the Keck 10-m telescope.

Down the rabbithole—the Keck 10-m telescope comes on line

In 1994 the first Keck 10-m telescope became available. This was to prove essential for breaking into the high-redshift Universe. The large increase in collecting area, about 8 times larger than the Canada–France–Hawaii 3.6-m telescope we had been using for spectroscopy, meant that for the first time we would be able to routinely identify spectra beyond redshift $z = 1$. The Keck observations would also prove critical for studies with the newly refurbished Hubble Space Telescope, now in its first year of sharpened vision, because for the first time we would be able to attach distances to the images of really faint, far away galaxies.

However, the startup of any new telescope has a shakedown period and, in the case of Keck, this was to prove a more unusual time than most. For one thing, the size of the instruments scaled up to match the collecting area of the 10-m diameter mirror. The effect was like sampling the bottle labeled "Drink Me" in Alice's trip to Wonderland—the filters and gratings that could be held in the palm of your hand for the largest telescopes previously used were now much bigger and heavier to move around, but still had to be positioned extremely precisely.

Secondly, the large mirror was composed of 36 hexagonal segments, and the mirror's shape held in place by a computer-run active control system, which was another new operational system.

The instrument we hoped would prove our case was the Low Resolution Imager and Spectrograph (LRIS) built by Judy Cohen and Bev Oke at Caltech. LRIS is a very sensitive instrument which provides high-quality images and spectra, and ultimately was to provide nearly all the spectroscopic information we have on high-redshift galaxies up to the present time. But the first observing night for this program ended abruptly when an engineering test on the telescope froze systems about an hour after startup. A second try months later stalled out three weeks before the assigned date, when Keck Observatory personnel decided to send the LRIS instrument back to refit it with more powerful motors. Finally, a 6-hour long Keck spectrum of these objects showed there was indeed no sign of any quasar-like features, and that these were likely to be redshift 4.55 galaxies viewed in the light of their star-formation stage.

The next step was to show that these emission-line galaxies could be found without targeting quasar fields. To launch this search, we would use the Hawaii Galaxy Survey fields and a set of custom-designed narrow-band filters for Keck.

Building our better mousetrap

The notion of using narrow-band filters on Keck had floated in the backs of our minds during this program, but we assumed that it would take a large and wealthy research consortium, or the Keck Observatory itself, to make this kind of investment. In the meantime, parallel scientific developments made us think about this again. First, deep HST images showed progressive changes in the appearance of the fainter galaxies—they appeared more irregular and fragmented. Redshift studies of these HST fields, particularly the extensive spectroscopic studies of the Hawaii Survey Fields, gave us galaxy distances. These data combined with ground-based infrared surveys that probed the peak of the light showed that galaxies were indeed smaller and fainter in the past. This "down-sizing" also explained the failure of earlier Lyman alpha searches, which assumed that very distant galaxies were as massive as their modern-day counterparts, and so predicted more expected emission from star formation. In another step forward, Chuck Steidel of Caltech and his collaborators succeeded in identifying a number of galaxies around redshift 3, which were again confirmed with LRIS spectra taken on the Keck telescope. Both Steidel's study and our own observations made it clear that really distant galaxies would be very faint, and extremely difficult to study without a bright feature like the Lyman alpha emission line.

Another difficulty in studying very high-redshift galaxies is that the reddest wavelengths observable from the ground are contaminated by atmospheric airglow lines, so highly redshifted objects are often viewed against a bright background. The rapid rise in both the strength and incidence of strong airglow emission lines at longer wavelengths means that even strong emission lines like Lyman alpha are readily detected in only two or three clear "windows" for galaxies above redshift $z \sim 4.5$. The best windows of low-atmospheric background correspond to Lyman alpha emission at redshifts $z \sim 4.5$, $z \sim 5.7$, and $z \sim 6.5$. Each of these windows can easily be covered with a narrow-band filter.

The new narrow-band searches would be a way to follow normal star-forming galaxies out to ever higher redshifts. To push these searches out to redshifts $z > 6$ we needed to learn the properties of these star-forming emitters: their typical brightness and how their numbers evolve as we look further back in time. In other words, we had to extend our cosmic census. Most importantly, since the emission line would prove key to getting to the very highest-redshift populations, we had to show that we could distinguish the Lyman alpha line from other emission lines of lower-redshift galaxies. The starting point for these studies should be the highest redshifts where other techniques could be used independently to identify high-redshift candidates, and later studies should work outwards in redshift. This put our first narrow-band filter at the modest Lyman alpha redshift of $z \sim 3.4$, where a number of galaxies were being identified by Steidel's group, and set our sample fields to be the very well-studied regions of the Hawaii Galaxy Survey, where a fairly complete census with Keck spectra was available. To this we would add the Hubble Deep Field (HDF), the target for a dedicated program of continuous, deep HST observations in four optical color bands for nearly two weeks in December of 1995, and the subject of a new but rapidly growing set of spectroscopic studies.

Our earlier paper identifying redshift $z = 4.55$ galaxies around a quasar had shown a 20-hour long narrow-band exposure on the University of Hawaii 2.2-m telescope to accompany the Keck spectra (see Figure 2.1 again). During the editorial correspondence over this paper I'd written that ". . . we are unlikely to see another Lyman-alpha discovery picture of the depth and image quality of [Figure 2.1] any time soon . . . [even] from Keck." I was now about to set out to prove myself wrong in this statement.

The larger instruments and optics of the Keck telescope meant that we would have to construct a large narrow-band filter for this project, while maintaining a high optical precision in its manufacture. At the time we proposed to do this, no narrow-band filters as large as we would need (9.5 inches × 9.5 inches) had ever been built to such specifications, and typical quoted estimates ran around \$25,000–\$35,000 for a single filter. Barr Associates, Inc., who had supplied our earlier filters, agreed to try to make a prototype for under \$6,000, and a Seed Money Grant from the University of Hawaii Research Corporation helped to fund this.

A long consultative effort with Boris Shnapir at Barr would finally pay off, but it was often nerve-wracking. Delivery of glass of the necessary size and quality for the filters ran months late, and then one of the pieces was discovered to be flawed. Could we rework the design using clear glass for the second piece? As the time for completion drew nearer, more problems arose. Because we were running late, even beyond a three-month safety margin that had been built in around the spring 1997 observing schedule, we negotiated a swap of our March Keck observing time for a May slot.

Although the prime target fields were no longer observable for the whole night, we would be able to use the filter before the current observing semester ended. During the coating process, which produces the good narrow response at the wavelength of the emission line, a call came in to report difficulties securing the filter in the coating tank; it was a little too big, could we cut down the square corners to fit it in?

There had been a maximum thickness specification, so that the filter would not protrude beyond the holder grabbed by the positioning arm. Could we relax this by changing the design for securing the filter in the holder, and use the extra thickness occupied by the retaining bar? The current filter was within the optical flatness specifications, but guaranteeing that a thinner version matched these tolerances would require an extra week at the polisher's plus about a week's worth of additional time to ship the filter back for optical verification and then forward it to the final delivery point.

The filter arrived at the summit about three days before the scheduled observing was to begin. The linear dimensions fitted the engineering drawing specifications, but were in fact a "pressure fit" to the inner dimension of the holder. Bill Mason, in charge of the Keck technical crew at the Summit, as well as a patient intermediary in the many technical discussions that ran fast and thick towards the end, took care of this by milling the inside of the holder.

The "excitement" of the filter fabrication was followed by an equally strenuous observing run. Probably in most cases, the "Aha!" of discovery doesn't come in the middle of the data taking, but only later during the calm moments of processing. Because our scheduled nights had been moved so far back, there would be no opportunities for a second try this semester. The usual process of designing spectroscopic masks for multiple objects is time-consuming. In the current instance, we would have to reduce the new-format observations as we took them, complete the analysis against existing data on the first night to identify sources, and then run the mask design software to set up masks that would be milled and loaded for the confirming spectroscopy in the succeeding nights.

Looking back . . . and looking ahead

The figure of Janus faces both ways at the threshold of each new beginning, and regards both past and future. The original redshift $z \sim 3.4$ searches demonstrated that high-redshift galaxies with strong Lyman alpha emission were commonly found, and that this was likely to be a good way of finding higher-redshift galaxies. A second filter corresponding to $z \sim 4.55$ was added later in 1997, followed by one which probed $z \sim 5.7$ in 1998 (see Figure 2.2, color section). These samples showed that there is far more star-forming activity taking place at early times than had been suspected. Since then, a number of galaxies at redshifts $z > 5$ have been found. As I write, we have just used this technique to find a galaxy at $z = 6.55$.

Looking ahead, a new generation of ground-based 8–10-m telescopes are poised to carry out these searches. Plans are underway for an 8-m Next Generation Space Telescope (NGST) which would have the collecting area to study faint, high-redshift galaxies and eliminate atmospheric airglow. This mission might take us to the epoch of first galaxy formation, around $z \sim 20$!

Suggested reading

Ron Cowen, Searching for the First Light, *Science News*, Vol. 153 (2 May 1998), p. 280

Richard Ellis, The Formation and Evolution of Galaxies, *Nature*, Vol. 395 (1 October 1998), p. A3 (Supplement)

Anne Finkbeiner, Seeing the Universe's Red Dawn. *Science*, Vol. 282 (16 Oct. 1998), p. 392

3

So we've lost the mission?
The Big Bang and the Cosmic
Background Explorer

JOHN MATHER, NASA Goddard Space Flight Center

John Mather is a Senior Astrophysicist in the Infrared Astrophysics Branch at NASA/Goddard Space Flight Center. His research centers on infrared astronomy and cosmology. He is the recipient of many awards, including the National Air and Space Museum Trophy, the American Institute of Aeronautics and Astronautics Space Science Award, the Aviation Week and Space Technology laurels, the Heineman Prize of the American Astronomical Society, the John Scott Award from the city of Philadelphia, the Rumford Prize of the American Academy of Arts and Sciences, and the Benjamin Franklin Medal in Physics from the Franklin Institute. In his spare time, John likes to read, listen to music, travel, and go to the ballet with his wife, Jane, a ballet teacher. John is presently working on several advanced space astronomy mission concepts, including the successor to the Hubble Space Telescope. Here he tells us of how he came to be one of the key players in NASA's COBE (pronounced, CO-BEE) mission to explore the Big Bang.

Two days after the Cosmic Background Explorer (COBE) satellite was launched, my wife heard me answer a 4:00 am phone call with the words "So we've lost the mission?". COBE had lost a gyro and we didn't know how well we would recover. Needless to say I got up, only an hour after getting home, to see what could be done. Fortunately, all would be well, and only a few weeks later, our team announced to the American Astronomical Society that the Big Bang theory was in good shape too.

We had proposed the COBE back in 1974, when I was only a few months out of graduate school, and now, 15 years later, all our dreams had gone into space, riding a rocket on a pillar of smoke. I will tell the tale of how the project got started, how a country nerd ended up standing in a dark field by the seacoast at dawn watching the launch, how it affected my life, and how it opened a new field of astronomy.

We built the COBE to look at the beginning of the Universe. Until 1929, only theologians and philosophers thought that the Universe even had a beginning. Scientists had almost no evidence. That year, the year of the great stock market crash, was also the year that Edwin Hubble discovered that distant galaxies are receding from us. Not only are they receding but, the farther away they are, the faster they are going, in exactly the pattern they would have if they were all debris from some cosmic explosion. Einstein had said it was impossible, believing without any observational evidence that the Universe could not be expanding, and told the Belgian abbot and scientist Georges Lemaître that "your calculations are correct but your physics is abominable." Einstein had previously introduced an extra term, the famous Λ constant, in his general relativity formulas to allow the Universe to be stationary. Nevertheless, the Universe does expand, and Einstein later admitted that his mistaken belief was the greatest mistake of his career. Curiously enough, the Λ constant is back in favor as a cause of an early rapid expansion, and it may be causing the expansion to accelerate even today.

Then, science took a break from fundamental research, and went to work in service of war, inventing radar, jet aircraft, and atomic bombs. In the late 1940s, as the world recovered, Ralph Alpher was a graduate student, and Robert Herman had just received his PhD. They were working with astronomer George Gamow to think about the early Universe. The three predicted that the expanding Universe must have had an extremely hot beginning, and computed the amount of hydrogen and helium that should have been produced by nuclear reactions in the primordial soup. They also predicted that the Universe should be filled by the residual heat radiation of that time, now reduced to a temperature of a few degrees above absolute zero. This radiation, now called the cosmic microwave background radiation, would be recognizable because it should come to us with the same brightness from every direction. It would have been difficult or impossible to measure with 1940s technology, even though it was predicted to be as bright as starlight.

By 1950, the debate about the nature of the Universe was very public, and the British Broadcasting Corporation (BBC) carried it live. Fred Hoyle, putting down the idea of the hot beginning, called it the "Big Bang," with full British innuendo, but the name stuck. Not much could be done to test the idea at first. The theory was worked out more completely, but the real breakthrough happened in 1965. Arno Penzias and Bob Wilson, working at the Bell Telephone Laboratories in New Jersey, discovered the microwave radiation, as they tested out some new receivers for the Telstar communications satellite. There it was, loud and clear, and suddenly the world of cosmology was different. Hoyle's Steady State Theory had failed to predict the radiation, and the Big Bang theory reigned supreme.

Curiously enough, Penzias and Wilson hadn't read George Gamow's popular books about the Big Bang, they hadn't read the original 1940s papers by Alpher, Herman, and Gamow, and they didn't know what they had found. Just down the road in Princeton, four people were looking for the radiation on purpose, and when they heard about the Penzias and Wilson results they immediately knew what they meant. They confirmed the discovery quite soon, but they hadn't read the books and old papers either. There's quite a tale there, of pride, social status, and credit for discovery. In retrospect we know of many missed opportunities going back to the 1940s. Penzias and Wilson got the Nobel Prize.

Growing up nerdish

I started out as a child, as Bill Cosby said. Back in 1953, Mars was very close to Earth, I was 7 years old, and my parents took my sister and me to the Museum of Natural History in New York City. We saw the giant meteorite at the Hayden Planetarium, we saw the model on the ceiling with the planets circling the Sun, we heard about canals on Mars, we saw the dinosaur bones stamping their feet, we saw the evolution displays of fish and human ancestors, and I was hooked. I wanted to know how we got here, from the beginning. My father studied dairy cattle breeding and feeding at the Rutgers experiment station in Sussex, New Jersey, and he told me bedtime stories about cells and genes. My mother was a grade school teacher and she read out loud from biographies of Darwin and Galileo. Her father was a bacteriologist at Abbott Laboratories, and had helped develop penicillin. Scientists were heros, and sometimes in great danger. I read Paul deKruif's *Microbe Hunters*, and thought about making the world a better place through science. I had nightmares about being imprisoned for my beliefs, or for teaching evolution in the schools.

I was only 11 when the Sputnik went up. Americans were already afraid of the Russians, and now we were desperately afraid. We had air raid drills in school, and were taught how to put our heads down under our desks. My father got a Geiger counter to find out if things were radioactive, and was part of the Civil Defense system. Suddenly it was good to be good at science and math. I got books every two weeks from the Bookmobile, which the county library sent around to farms. Even the library itself was brand new. We had a science fair, and I saved up my allowance, a quarter a week for a long time, to buy a Heathkit short-wave radio with five vacuum tubes. I put it together myself, but it didn't work because my soldering iron was meant for roofing, and had melted some parts. A few months later I found out how to get some new parts, and suddenly there were voices from far away. I studied the parts catalog from Allied Radio the way other kids memorized baseball statistics. I built a "robot" with some vacuum tubes and motors from my Erector set, and entered it in the science fair, but it didn't do anything. Transistors were invented, and *Boys Life*, the Boy Scouts' magazine, carried articles about how to build radios. Microwave relay towers were built on the mountain nearby, and one of the engineers there started up a 4H club for electronics.

By the time high school came around, the country was supporting summer schools for science kids. I learned math at Assumption College in Worcester, Massachusetts, one summer, in an old red-brick building whose cupola had been touched by a tornado while people were praying inside it. I learned physics at Cornell University the summer between my junior and senior years in high school, and now I thought I might really be able to be a scientist. I'd seen some laboratories and I loved the energy of my favorite teacher, Mike Nieto, who was a graduate student. I was even pretty good at the work. I got back to telescopes, saved up my allowance, and assembled a small reflector from parts from Edmund Scientific. I borrowed *The Amateur Telescope Maker*, all three volumes, from the library over and over. I tried to measure the motions of asteroids and compute their orbits, but the math was much too hard for me. I tried to learn it from a book, but Gauss, who invented this subject in the mid nineteenth century, was way ahead of me (and still is). I did enter this project in my high school science fair, and it went on to state level and won me a trip to Chicago and an invitation to go on a Navy cruise.

College and graduate school

College was quite a challenge. I went to Swarthmore, warned by my parents that I'd been a big fish in a little pond, and I would have to study very hard to win again. I did, and it worked. I was keenly aware that they were paying for me to go, and I was determined to get every bit out of it. From there it was off to Berkeley for graduate school. That was a much bigger pond, and a real shock. Swarthmore was a little school, only 1,200 people in a small town. Berkeley was huge and at least the physics students weren't very social. They'd come in to class and sit down with their books and read.

The psychology students went to class and planned their adventures and their parties. After a couple of years of taking classes and going to the library, I was fairly tired of school. Then came my lucky break. It was time to find a research topic and a research professor, and I met some wonderful mentors. Paul Richards was my thesis advisor, and in his laboratories I worked on designs for instruments to measure the cosmic microwave radiation. Mike Werner had just received his PhD and was working in Charles Townes's group, and they taught me a lot too. It was 1970, just five years after the radiation had been found, and the news from a rocket experiment said that the Big Bang theory wasn't right. The radiation was 50 times too bright. Worse yet, a mountaintop experiment said that there was a spectrum line in the cosmic background radiation, a frequency where the radiation was much brighter than at nearby frequencies. The Big Bang couldn't do that, so maybe the radiation wasn't cosmic after all. We ought to check.

It took us a long time. First, we built a new instrument to take to White Mountain in California. It was called a Fabry Perot interferometer and it was really tricky, especially for our first effort. I worked with Mike Werner on this project. We helicoptered ourselves and the apparatus up the mountain in the winter and tried to breathe. At first our fingers and tongues were blue from lack of oxygen, but after a few days the headaches went away and our color came back and we could think a little again. After two trips we concluded there was nothing wrong with the Big Bang radiation that we could see. Alas, our ability to measure the cosmic radiation was limited by the air overhead, which emits its own radiation.

Our next adventure was to Palestine, Texas, a small town south of Dallas where scientific balloons are launched. Our new apparatus hung by a thousand-foot cord from a huge polyethylene bag, as big as a football field. It would do a better job than we could manage from the mountain, because it would go above 99.5% of the air. This new project took until 1973 to get ready. We got impatient. More tests would take a long time, and they wouldn't be very realistic. Maybe the apparatus would work. We (my fellow graduate student David Woody and I) drove it to Texas on a yellow University truck, across the Arizona and New Mexican deserts to the lush greenery of watermelon fields of East Texas. We launched it, or I should say a lot of people launched it. The crew to handle these huge things is very professional and they have the most amazing equipment. Tiny Tim, a converted Earth mover, dangles the payload from his huge jaws 20 feet in the air, while the balloon bag rises overhead, and then races across the field with it until the cable pulls tight and the balloon lifts our work into the sky.

Well, it didn't work. It didn't work for three different reasons, which we found out after we got back. That night was awful. Three years of work went up, up, and away, and there wasn't a thing we could do about it. We sat in the control room, thinking about what to do to recover, and sending computer commands, but nothing helped. It was a defining moment. I decided that my Zen needed revision. I couldn't, I wouldn't ever, be so impatient. I would test everything. This time, Paul let me finish my thesis on the basis of the previous work, and in January 1974 I left California for a new life. David rebuilt the apparatus and flew it again three times after I left, and it worked twice. The measurements said the radiation had just the right spectrum to match the predictions, and the Big Bang theory was still OK.

Going to work for NASA

I would be a radio astronomer. Pat Thaddeus in New York City, at NASA's Goddard Institute for Space Studies, had just built a new telescope on the roof of the physics building at Columbia University. I wanted out of cosmology. I wanted to do something where it didn't take years to build the apparatus and then see it fail. Pat got me started observing with a radio telescope and making some computer calculations, and I even made a little progress. However, the fates had something else in mind: NASA. NASA had sent around a team to Berkeley to see what their Space Sciences Laboratory there was doing, and I had told them about our balloon project. They wanted to know why we weren't doing it in space. I thought, "Who, me, I'm just a kid?"

In summer of 1974, NASA issued a nationwide call for satellite proposals. Pat said we should all think of ideas. There was only one thing I knew anything about, my ill-fated thesis experiment. By now the emotional sore spots had worn off and I thought maybe it was worth doing in space. It could be done thousands of times better than we could imagine doing even with a balloon. Pat said I should call his friends and assemble a team, so I did. Six of us wrote a very thin proposal for the "Cosmological Background Radiation Satellite," and sent it in. We wanted to build four instruments, three of them inside a tank of liquid helium, and put them in space.

We had three objectives. First, we would measure the spectrum of the cosmic background radiation a thousand times better than we had done with my thesis experiment, and compare it directly with a nearly perfect blackbody. A blackbody is an object that absorbs all radiation that falls on it, and it is also a perfect radiator whose brightness follows a simple formula. If the Big Bang theory is right, the background should match a blackbody at a particular temperature, which we would measure. Second, we would look to see if the microwave radiation is equally bright in all directions, as it should be if it comes from the Big

Bang, and then we would look for little hot and cold spots that might be the seeds for galaxies and clusters of galaxies. Third, we would look for the light from the first galaxies. Maybe the early universe is filled with galaxies that are too far away for any telescope to see them, but we might still find the hazy glow.

I drew a picture and a draftsman tidied it up (this was before computers could draw). In retrospect it amazes me that so much could come from such a little booklet. Now, in today's intensely competitive environment, such a short proposal would have no chance, but in those days most proposals were about as thin as ours. I have to think we had a guardian angel, and it was true, we did: Nancy Boggess was at NASA Headquarters, and she was a strong advocate of the new field of space infrared astronomy. Also, major scientific advisory committees had told NASA that our subject was very important.

In reality, though, our fate was to compete with over a hundred other proposals. Two other groups had put in ideas related to ours, one from Berkeley, and one from the Jet Propulsion Laboratory in Pasadena, California. At first, NASA thought one of our instruments (the one most like my thesis experiment) might go along with another mission that wanted a helium cryostat, but that turned out to be much too difficult. Instead, NASA formed a new team from members of our group and the Berkeley and JPL teams. We would figure out what to do now. In 1976, I took a job at NASA's main science laboratory, Goddard Space Flight Center in Greenbelt, Maryland, in hopes that our new project might become real. If the project were selected, I would be its lead NASA scientist, and I would be in charge of one of the instruments. Suddenly I was the center of a whirlwind. Be careful what you ask for, you might get it! I was 30 but I still felt like a kid, a bit awkward with words, and when I had to give a speech for the first time I got cold sweat running down my back. Maybe it was a good thing that I didn't know to be afraid of what I was getting into.

Now we had a chance. NASA sent the twelve winners of the first round of competition a little money to support writing a more complete proposal. We sent our bit out to our team members and to Ball Brothers in Boulder. Ball Brothers spent a lot of their own money too, in hopes of winning some contracts when the competition was over. We wrote a very thick proposal this time, two volumes each an inch thick. It demonstrated we could do this mission within the allowed budget, and it would measure the Big Bang radiation and look for the radiation from the first galaxies. We even decided on a new name, the "Cosmic Background Explorer," or COBE. Review committees smiled upon it. NASA gets external advice from scientists around the country, and they apparently felt that the obvious difficulty of the work was still acceptable because of the tremendous importance of the results we might get.

Building a team

So Goddard Space Flight Center built us a team (Figure 3.1). The International Ultraviolet Explorer was just getting completed (it operated successfully for 18 years before it was turned off), and their management took us under their wing. They knew how to do things, and they had a working organization. Would we build the equipment at Goddard, or would we buy it? I was nervous about buying it, because nobody knew what to buy. Nobody had ever designed anything like what we wanted. Ball Brothers were good, but they hadn't built these instruments either, and they were getting expensive as they realized what it would take to do the job. They were building a cryostat (the liquid helium tank we would need) for another project, so we planned to buy another one from them.

COBE's instruments would be advanced so far beyond what anyone had done that many good engineers thought it was impossible. They were almost right. We didn't know how to do it at Goddard either, but at least the scientists and engineers could work together there. If we bought from a big aerospace firm, I was afraid we'd be talking to lawyers and accountants instead. In the end though it wasn't a matter of principle, it was cost. Goddard adopted us. It would contribute manpower, which wouldn't come out of the budget. The engineers wanted something so challenging that they could use it to attract good new talent. There was just one string attached. Other projects had priority. If company X made a big mess of the work NASA was paying for, NASA had to pick up the pieces and make things right.

So we fought, politely. Our team wanted the best engineers, but so did

Figure 3.1   The COBE Science Team (John Mather is just right of center in the back row).

everybody else. Our team wanted priority in the shop. So did they. Worse than that, Goddard (like many engineering companies) has what's called a "matrix organization," in which everyone has multiple bosses who argue over who works on what. The matrix organization would be the death of us as we tried to claim our percentages of time from each person. It wasn't working. The only thing that broke the logjam was a national disaster. In January 1986, the Challenger exploded. Nothing would bring back the dead astronauts. Other rockets exploded in the ensuing months, both American and European. Things looked extremely grim everywhere. National pride stepped in, and people refused to let NASA die too. Congress gave money, and NASA would build another Shuttle. But what would COBE do in the meantime? COBE was going to ride on a Shuttle, and so was practically every other NASA payload. That was the bargain with the White House and the Congress. So we were stuck.

Recovering from disaster

Dennis McCarthy, our Deputy Project Manager, found a way. He talked to other countries about partnerships for COBE, in which they could provide a rocket to go with Goddard's spacecraft. NASA Headquarters heard about it and threatened terrible things. NASA would have to find a way to launch COBE with American rockets. How could American pride be maintained if COBE went on a foreign rocket? The very thought was appalling. Dennis found the way. There were parts for an American Delta rocket, and there might be enough to build a whole one. The COBE, which weighed 10,500 pounds loaded with fuel, might be shrunken to 5,000 pounds, the maximum the Delta could put in our orbit. In a few months, we had a plan. We would launch in early 1989, and we would have top priority. We would be NASA's first science mission after Challenger, and America would be proud. We would make a "skunk works," named after the famous Lockheed facility where spy planes were built for the Cold War, and we would bring together the key team members in one place. Nobody could stop us now, and we could insist on immediate results, and overtime (lots of it).

Needless to say, two years was a short time to finish the project when we had to build a whole new spacecraft! We built two of them to make it faster, one to be tested on the shaker and one to fly. The shaker could make 35,000 pounds of force to simulate the launch, and it was a frightening sight. We worked nights and weekends most of the time. Families wondered where we were and when we would ever be done. Vacations were deferred, sometimes for years. We had to keep the instruments extremely clean, so we would be sure we were seeing the beginning of the Universe and not just dirt on the mirrors. Some of us spent months in the clean rooms wearing white bunny suits with masks and gloves. We built a "car wash" to clean the parts, and we had several people there round the clock to do it. We no longer had time to make things better. We just did what we had to do, and we only fixed things if we had to. Voltaire said "the better is the enemy of the good," and our team believed it. Even scientists like me had to give up some cherished hopes. Better detectors, more calibration tests, more software and computers, we gave them up.

One night I woke up in a cold sweat. I had just realized that I had designed a fatal flaw in the calibrator for the spectrometer, the instrument for which I was responsible. The next morning I called for help, and after careful calculation we found a solution. It needed more thermal blankets, and it would be cold enough after all. We put the spectrometer together, and it worked. Then we put in a better mirror mechanism, and this time it didn't work. A tiger team was formed, and we found out what we had done wrong. It took just a year to build a new one and put it in.

After we had the whole payload together, in spring of 1989, our engineers insisted on a new test, one in a different orientation. We can't simulate zero gravity on the ground, but we can set up the equipment so gravity has the least possible effect. This time, we tested the calibrator by making its pivot axis vertical, so gravity wouldn't make it swing. The calibrator failed the test. It wouldn't stay in place without the help of gravity, and it wouldn't work in space. We pretty quickly knew what was wrong, and we were lucky. We could fix it without taking everything apart. Another few months and we were ready to ship the payload to California.

It rode down the Capital Beltway on a special, very slow moving big truck at dawn, and went to Andrews Air Force Base. There, the truck drove onto a giant C5-A aircraft, and flew all the way to California's Vandenberg launch site. There, the truck drove back off, the COBE parts were tested again and reassembled, and the whole thing readied for the top of the Delta rocket. Everything seemed OK. Then, the October 1989 earthquake came, the one that leveled highway bridges in San Francisco. We were only hundreds of miles down the coast, and the payload might have been hurt, but luck smiled on us. The delicate mirror mechanism was safely bolted down that day because the two engineers who might have been testing it had taken that day off to get married.

Launch!

Finally came the readiness reviews. Was the rocket ready? The parts had been brought back from the graveyard because the Delta production line had stopped years ago, and some fuel tanks had to be patched where pigeon droppings had eaten holes through them. We heard that a wet rag had been left in a pipe during a welding operation, but the pipe was tested and the rag was retrieved. It's impossible not to make mistakes, so it's essential to catch them and fix them. One might say luck was with us, but this one wasn't luck, it was a test procedure based on long hard experience. Of the nearly two hundred Delta rockets before ours, only four failed. On the last day before launch, the rocket guidance computer had to be replaced. Were we supposed to take this as good luck or bad luck? How could we know? Everything was as ready as we knew how to make it, but we all knew that there was no way to tell if it would work. What about all the mistakes we didn't catch? The only way to know was to push the button.

So that's how we came to have 1,500 people standing in the fields around the launch site in Lompoc, California at dawn on November 18, 1989. In the daytime, the flat spots were beautiful with commercial flower growing, and the hills were steep and covered with grass and live oaks. In the early morning, before dawn, it was cold and dark and windy, and we shivered. The balloons were sent up to find out about the wind above us. At first the wind was too strong, but then it slowed down just enough. A strong wind would blow the rocket off course, and when the rocket nozzles swiveled to compensate, the sideways forces could destroy the rocket. When the button was finally pushed, we were miles from the rocket, and the light came to us long before the sound. At first slowly, then faster and faster, the pinpoint of light climbed to the sky and disappeared. The wind wound the exhaust trail into a pretzel shape near the Moon, the rising Sun lit it up, and it was spectacular. The wind had almost destroyed the rocket, but not quite.

Parties, champagne, crises, and science

Now was the time for parties and champagne, and for heading back to Goddard to run the spacecraft and turn on the instruments. Within minutes the rocket was out of range, and we wouldn't hear anything from it until it came in range of the ground stations on the other side of the world. Each orbit took 103 minutes. Back at Goddard the next day, I learned that all had worked well, everything was as expected. Then the gyro failed. Fortunately, we had six, and we needed only three. The spacecraft was a bit wobbly, but it was alive, so we learned how to run with a dead one and continued on.

Just a few weeks later, on the second Saturday in January in 1990, we presented our first results to the American Astronomical Society in a giant hotel ballroom in Alexandria, Virginia, near National Airport. I was worried that our event was on Saturday, and thought everyone would have gone home already. All of us were totally exhausted, having stayed up to all hours to make the instruments work and learn what they were showing us. I was amazed to see a packed auditorium, well over a thousand astronomers. Nancy Boggess, who had backed us for years at NASA Headquarters and was now working at Goddard with me, gave an introduction. Mike Hauser and George Smoot, the lead scientists for the other two instruments, gave their talks, and I gave mine. I showed a spectrum from the Far Infrared Absolute Spectrophotometer instrument (FIRAS). The plot showed the brightness of the cosmic microwave background radiation at all wavelengths from 0.5 to 5 mm, and it matched the theoretical prediction exactly. I said very little, just projected the plot on the screen, and there was a standing ovation. Everyone knew what it meant, and why it was so important. I was absolutely unprepared for such an outpouring from my colleagues, and could barely say that it was now time for the next speaker. The Big Bang theory had withstood a great test, and it was fine. The COBE team had also withstood a great test, and we were fine.

We found that the cosmic background radiation has a temperature of 2.735 ± 0.06 kelvin, just a little above absolute zero, and that the difference between the measured spectrum and the perfect blackbody was less than 1% (Figure 3.2). Nothing anyone could imagine but the Big Bang could make such perfection, so the Steady State theory now had a stake through its heart, despite the persistent efforts of Fred Hoyle and his colleagues to resurrect it by improving it. Eight years later, after all the data were analyzed and calibrated, we could say the radiation is even more perfect: 2.725 ± 0.002 K, and only 5 parts in 100,000 difference between the cosmos and the perfect blackbody. The answer was 20 times better than we had dared to hope.

Two years later in April, we announced another breakthrough. Our second instrument, the Differential Microwave Radiometer (DMR) had mapped the sky, looking for hot and cold spots in the microwave radiation, that might give some clue about the Big Bang. Sure enough, they were there, but they were extremely faint. These hot and cold spots

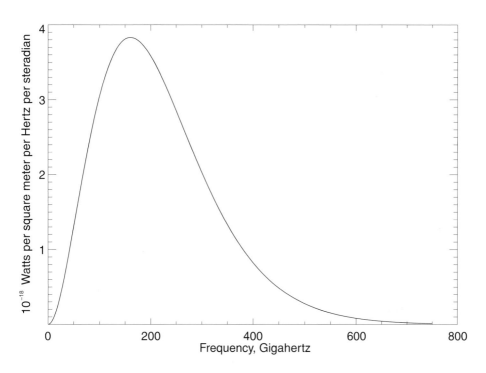

Figure 3.2   Spectrum of the 2.7 kelvin cosmic microwave background radiation. Theory and observations differ by less than a part in 10,000.

were only a part in 100,000 different from the average temperature. George Smoot, who was the lead scientist for this instrument, got a lot of publicity for saying it was like looking at the face of God. He wasn't the first to use that phrase, but it brought a huge wave of public attention, and controversy. Religious folks wanted us to agree that our results supported their versions of history. I was interviewed for a Catholic religious television channel, and our findings were written up in Japanese and Arabic, and reported around the world. George got an offer to write a book for a huge sum of money. The Vatican Observatory (yes, the Pope supports cosmology) held conferences, and Galileo was rehabilitated around then. Scientists wrote thousands of papers citing and interpreting our results, and the maps and the spectrum plot are now in virtually every astronomy textbook.

These hot and cold spots show the Universe as it was about 300,000 years after the Big Bang (Figure 3.3, color section). That's the moment when the hot gases of the Big Bang cooled down enough to become ordinary hydrogen and helium. Before then, the gas was ionized and opaque, and afterwards, the gas was transparent and the primordial heat radiation was free to go in a straight line. According to computer simulations, the hot and cold spots are responsible for our existence, because they were the primordial seeds around which galaxies and clusters of galaxies would grow. The search for more information about these seeds is still a very hot topic, with dozens of projects around the world, one space mission (the Microwave Anisotropy Probe, MAP) being built at Goddard for launch in 2001, and the Planck mission being planned in Europe for launch in 2007. With luck, we'll know how long ago the Big Bang happened, how much matter there was, of both normal and "dark" varieties, and whether the expansion is slowing down or speeding up.

Our third instrument, the Diffuse Infrared Background Experiment (DIRBE), finally yielded its secrets in 1998. Mike Hauser's team found the light from the first galaxies, and it's much brighter than most of us expected. Apparently more than half of the starlight from the early times was absorbed by tiny dust grains and converted into infrared radiation, so we would never have known about it with normal telescopes operating on the ground. This is one of the great surprises of science. Theorists had told us what to expect about the spectrum and the hot and cold spots, but they didn't tell us this one. Now, big telescopes on the ground are beginning to work at some of the wavelengths where these galaxies can be seen, and space missions are being planned to look at them without the interference of the atmosphere. A whole new domain of science is now open, and we know there's something important to find!

Some people think the end of science is near, but I don't. The world is dangerous. The threat of war, terrorism, plague, and natural disaster of all sorts is very strong, and people are investing in technology to protect themselves. Astronomers have benefited from generations of technological advances and, despite the end of the Cold War, there's no reason to think that will stop. There's also no end in sight for what computers can do for us, and I think they'll be a great help. Moore's Law says that computers double in speed and memory every year or two, so how long does it take before they are so powerful that they do things we would never dream of today? This isn't a government bureaucrat's project, it's the response of the marketplace to opportunity. (By the way, the Internet and the Web were invented by government scientists in the US and Europe, and then made public.) Maybe one year I'll be able to walk into my office and say to my computer, "Hey, Nerdina, I think Congress might be ready to approve of a new telescope 30 meters across. How would you build it?" I'm already working on the Next Generation Space Telescope (NGST; see Figure 3.4, color section), a successor to the Hubble Space Telescope that would be 8 meters across. The NGST would be capable of seeing those first galaxies that formed after the Big Bang and perhaps produced the infrared radiation that Mike Hauser found. I don't have Nerdina to help yet, but maybe next time . . .

Suggested reading

Alan H. Guth and Alan P. Lightman, *The Inflationary Universe: The Quest for a New Theory of Cosmic Origins*, Pegasus Press, 1998

Anne Kinney, When Galaxies Were Young, *Astronomy Magazine*, May 1998, p. 42

John C. Mather and John Boslough, *The Very First Light*, Basic Books, 1996

Martin J. Rees, *Before the Beginning*, HarperCollins, 1998

Hervey S. Stockman (ed.), *The Next Generation Space Telescope, Visiting a Time When Galaxies Were Young*, Space Telescope Science Institute, Baltimore, MD, 1997

4

Computational adventures in cosmology

NICK GNEDIN, University of Colorado

Nick Gnedin was raised in Russia. He received his Master's degree from the Leningrad (now St. Petersburg) Polytechnical Institute, and his PhD at Princeton. For two years he worked as a Research Assistant at the Astrophysical Department of the Ioffe Institute for Physics and Technology (Leningrad, USSR). In May 1991 he was invited to Princeton University, and has remained in the United States ever since. Most recently, Nick has been a professor in the Astrophysical and Planetary Sciences Department of the University of Colorado. He and his wife Marianna are raising their daughter Nina in the Rocky Mountains while Nick teaches and continues to pursue his research love—numerical simulations of the evolution of the Universe since its early youth.

The greatest happiness of the thinking man is to have fathomed what can be fathomed, and quietly to reverence what is unfathomable.

Goethe

Ever since an animal looked up to the night sky, wondered at the brilliance of stars and the vast depth of space, and in the act of doing so became a human being, the Universe beyond our immediate locale was always a subject of human curiosity.

What are we in this world, and how do we relate to the immense emptiness around us that we call space? How did the Universe come to existence? Was it born in a divine act of Creation, or has it existed forever?[1] Innumerable generations of philosophers, scientists, thinkers have asked those questions, just as we are asking them now, and as our distant descendants will be asking them in the ages to come. For no matter how hard mankind tries, or how clever future generations are, there will never be found all of the answers to those questions.

---

[1] A related question that a reader may amuse himself or herself with thinking about is whether an "existing-forever universe" can be created.

55

Perhaps those were the questions that pushed a young Russian boy to pursue a career in astronomy, or, perhaps, it was something else . . .

The beginning

The life of any person is like a road. Some people's road is a straight stretch of highway roaming through Great Plains. Others' is a mountain side road, making 180° turns every few hundred feet. My life is like an interstate highway on the east coast: it goes straight for quite a while, and then makes a wide turn. And whenever such a turn awaits, you need a road sign to warn and direct you. In human lives, such road signs are good people. In my life there are two people whom I will be grateful to until the end of my days for pointing me in the right direction at treacherous turns.

The first of these people is my father, Professor Yurii Gnedin, a well-established professional astronomer and the Associate Director of Pulkovo Observatory in St. Petersburg, Russia, who instigated and challenged me to risk a career in astronomy. Recently, I was going through a pile of very old greeting cards which I found at the bottom of a drawer in my parents' house, and was reminded of just how early my father's influence was felt. Among the dusty pile of angels and Christmas trees, I found a card written by my kindergarten teacher when I was entering the first grade. She wished me to achieve my ambition of becoming an astronomer. When I read it, I could not believe my eyes. At the age of six, you do not want to become an astronomer! You want to be a police officer, a bus driver, a sailor, an astronaut, perhaps, but an astronomer? What six-year old would want to change a life full of Earthly or cosmic adventures for a dull existence as a star-counter or a paper-worm?! Yet, there was a fact, complete with a written record and a valid signature. I did want to become an astronomer at the age of six, and I did become one. And for that I must thank my father, if a mere "thank you" can reciprocate for the very first lifelong decision that I ever made, and that was *right*.

So, having started my scientific career at the age of six (or so it seems to be), I continued it along the straight stretch of my highway, through

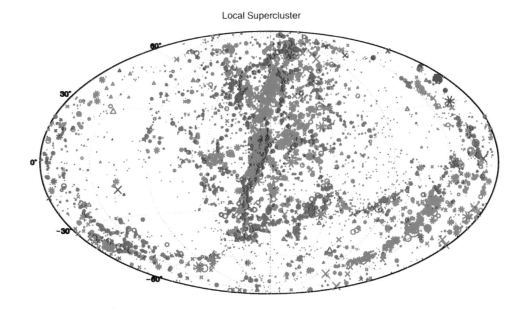

Local Supercluster

Figure 1.3. The current best map of the local supercluster (LSC) of galaxies. This shows a projection on the sky of all the known galaxies with measured redshifts less than 3,000 km/s (i.e., galaxies nearer than about 130 million light-years). In this plot, galaxies that are blueshifted are in blue, galaxies with apparent radial velocities between 0 and 1,000 km/s are in red, those between 1,000 and 2,000 are in magenta and those between 2,000 and 3,000 km/s are in green.

The local supercluster is the large band of galaxies running nearly vertical on this plot. The map is an equal area Aitoff projection of celestial coordinates. Our own galaxy, the Milky Way, lies at the edge of the "disk" of the local supercluster, much like the Sun lies near the edge of the disk of the Milky Way. The areas of the map that are only poorly covered are those parts of the sky essentially blanked out by dust in the Milky Way, through a glass, darkly.

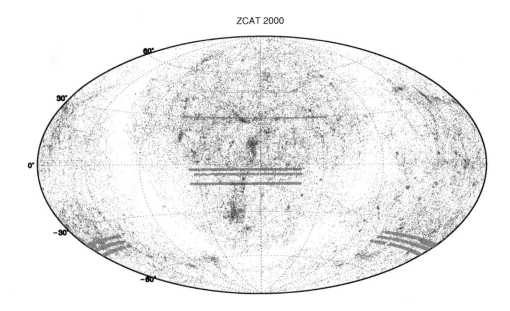

Figure 1.4. Plot of the distribution on the sky of all known galaxy redshifts. The sampling isn't great, as in Figure 1.2, but you can easily see the large areas of sky obscured by our own galaxy, and you can also see where astronomers have probed especially hard. Galaxies marked in blue are closer than about 30 million light-years, those in red are between 30 and 200 million light-years, those in green are between 200 and 500 million light-years, and those in cyan are out beyond that. A few really deep strip surveys, primarily the Las Campanas Redshift Survey, are seen as cyan strips.

Figure 2.2. A colour image of a galaxy at $z = 5.74$ found in the Hawaii Survey Field studies.

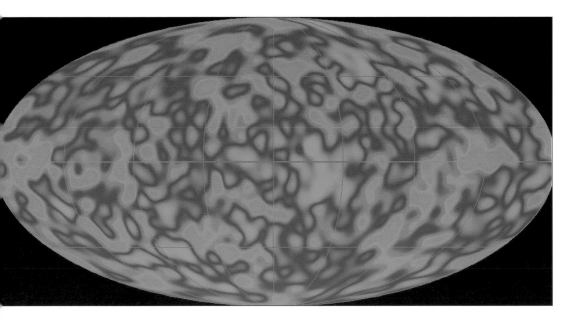

Figure 3.3. Hot and cold spots in the cosmic microwave background radiation, showing the Universe at an age of 300,000 years. Hot and cold spots differ from the average brightness by a part in 100,000. Courtesy of GSCF and the COBE Science Team.

Figure 3.4. One concept for the Next Generation Space Telescope, 8 m in diameter, to be launched around 2010 to see the first galaxies.

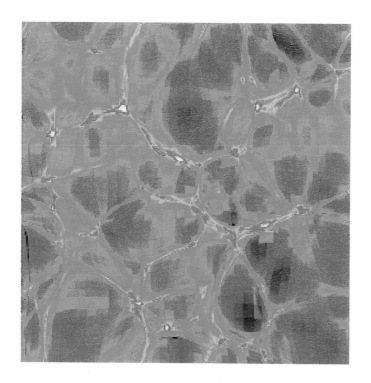

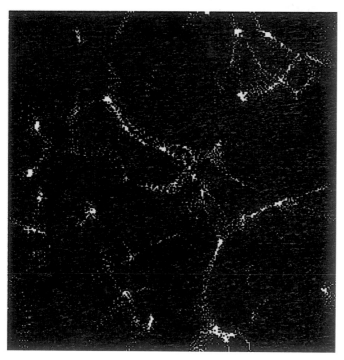

Figure 4.3. A comparison of the dark matter distribution (above) and the galaxy distribution (below) on large scales. From a simulation by the author (Nick Gnedin).

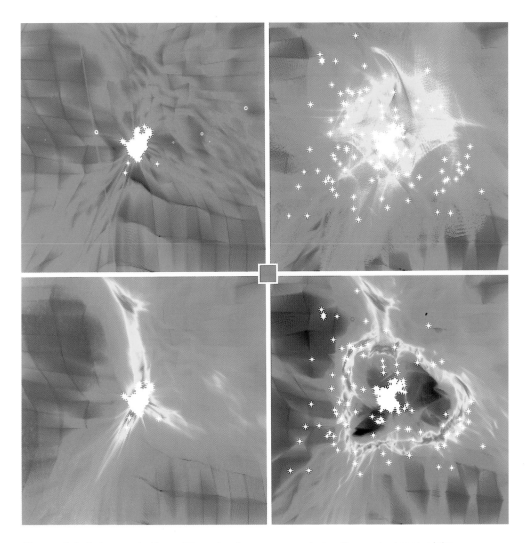

Figure 4.4. A demonstration of the role of supernovae during the early stages of the evolution of the Universe. The right column depicts the realistic Universe, and the left column shows the Universe where supernovae explosions are forbidden. On the top row is the temperature of the cosmic gas, and on the bottom row is the gas density. Stars are shown with white symbols.

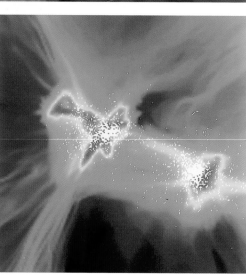

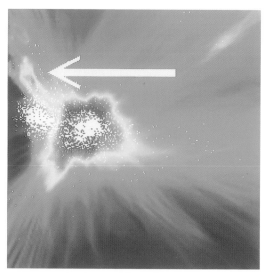

Figure 4.5. "Chipping" out gas during galaxy collisions; white dots are stars, and colors show the gas. The picture shows two small galaxies moving toward each other (the left panel), colliding (the middle panel), and chipping off gas (the right panel). The largest chip is pointed to by an arrow.

the elite high school and the Leningrad Polytechnical Institute (which soon after that changed its name for the more prestigious-sounding St. Petersburg Technical University) to the next (and, I hope, the last) drastic turn in my fate. It was the fall 1990, and I was a research assistant (which was a permanent, if very poorly paid, position in the academic system of the USSR) at a world-class research institution—the A.F. Ioffe Institute for Physics and Technology—in (then) Leningrad. The person who changed my life then and forever was Jerry Ostriker, then a chair of the Department of Astrophysical Sciences, and now a provost at Princeton University.

At that time Jerry was visiting the Soviet Union, and I had a chance to show him the work I was doing with Alex Starobinsky on numerical simulations in cosmology. Jerry was already the leading scientist and one of the cofounders of a very young field of numerical cosmology (and had he not been the leading scientist in a dozen other fields of astronomy and written a wonderful chapter on dark matter in this volume, it would be him, not me, writing this chapter). He had the wisdom and courage to invite me, an unknown Russian youth with a poor command of English, to spend a year at Princeton as a long-term visitor. I must confess, for him, that was indeed a gamble, since at that time much of my "research" experience was limited to trying to run, on a little PC with 640 kilobytes, simulations that others run on supercomputers with hundreds of megabytes of memory.

But, one way or another, I took the wide turn in my highway with Jerry's help, and after a nine-month fight with the ultra-powerful bureaucratic machine of the idiocy called the Soviet Union, I found myself in June of 1991 at the doorsteps of Peyton Hall, which houses the Department of Astrophysical Sciences of Princeton University. I crossed the threshold, fully intending to cross back again one year later, but fate delayed this moment by four more long and wonderful years.

So 1991 brought the largest change of my life, as I changed the country where I lived. But 1991 also changed the lives of another 400 or 500 million people: the Soviet Union, and the communist nightmare with it, ceased to exist. And so it happened that by the end of 1991 I found I had nowhere to go back to: the city I came from, Leningrad, became St. Petersburg, finally shedding the hateful name. And the country I came from, the Soviet Union, fell apart like a house of cards.

At this treacherous point, I came close to exiting my astronomical highway altogether in search of another road, and it was Jerry Ostriker again who guided me, by mildly pushing me into applying to the graduate school at Princeton.[2] And so I did, and my career took a down-turn: I went from a permanent position at the Ioffe Institute to a one-year post-doctoral type position at Princeton, and after that into the graduate school. Such a "downfall" can make a person dizzy, and I gratefully recall Jerry's help and concern during my years as a graduate student. With Jerry's help and my own efforts, I raced through my graduate years at twice the speed limit (thank God there wasn't a patrol car on the way!), finally getting my PhD in astrophysics in 1994 with my thesis on numerical cosmology, supervised by Jerry.

Two cosmologies

So what is it, this numerical cosmology, what is it about, and what do we need it for? We all live in a universe, and cosmology is a branch of astronomy and physics that studies this Universe. Or, more precisely, there are two cosmologies: physical cosmology, which is mainly concerned with physical processes in the early Universe, how the Universe evolved when it was very young, and what we can learn about this early stage in the life of the Universe through astronomical and physical measurements; and "astronomical cosmology," usually called extra-galactic astronomy. Astronomical cosmology is much closer to home. It is the field of astronomy that studies how the astronomical objects we see in the sky came to exist, how they evolve, and what physical processes are taking place during various stages of the recent evolution of the Universe.

The main difference between those two cosmologies is that, at least within the framework of the standard Big Bang theory, the Universe was homogeneous (very smooth) on all relevant scales in its early history. However, it is far from smooth right now: we see galaxies, which themselves gather into galaxy groups and clusters, and the clusters in turn

---

[2] At that time I did not have a PhD, so that was the only possible way for me to continue my scientific career.

form superclusters on very large scales. Physical cosmology studies the Universe when it was smooth, and thus simple. Essentially all the properties can be computed very accurately with a modest expense on computer resources, and these predictions can then be compared with whatever limited volume of observational data we have on the early Universe.

The situation is very different with extragalactic astronomy. First, the amount of the observational data available is enormous, as there is a large number of galaxies (including our own) which are studied in extraordinary detail, and a much-much larger set of galaxies for which some data (but not a complete picture) exist. In addition to that, there is the Intergalactic Medium (usually abbreviated to IGM), the cosmic gas filling the space between the galaxies, and it may even be better studied observationally than the galaxies themselves.[3]

So by no means does extragalactic astronomy lack observational data! It seems that life should be good for cosmologists studying the recent Universe: the data are so abundant, just go and model the Universe, and you will be able to define your model very precisely because you can test and refine it against so many data! It is those scientists who work in physical cosmology who are concerned with obtaining the data,[4] it is they who keep each tiny piece of observational information, every number obtained from the real data as a priceless piece of jewelry, looking at it from every possible angle and using it 110%.

This is precisely the difference between physical cosmology and extragalactic astronomy ("astronomical cosmology"). If physical cosmology is granted only a very limited amount of observational data, its theory is well developed because it is relatively simple and fully computable with existing computers. Extragalactic astronomy enjoys a vast amount of existing data (and these data grow at an ever-increasing rate), but at the same time it lacks well-developed theory, simply because

---

[3] Mostly because a parcel of gas is a much simpler physical entity than a galaxy, so it is possible to proceed much further in understanding this parcel of gas, compared with understanding of an external galaxy, given roughly the same number of observational resources.

[4] The early Universe is too far back in time, and what happened then was later covered by "layers" of more recent physical phenomena. Physical cosmologists are like archeologists who have to dig deep into the ground to find scarce pieces of times past.

the Universe of extragalactic astronomy is not simple and smooth any more, it has complicated and diverse structures on a large range of scales, and incorporates the whole zoo of cosmic phenomena. It is this garden variety that strongly limits our ability to predict and even describe in physical terms what is going on with galaxies and the IGM today and a few billion years ago. It is because of this complexity that *the only possible way to devise a comprehensive theoretical description of the Universe is by using numerical simulations on supercomputers.*

Of course, one is still allowed to introduce simple models based on some *ad hoc* assumptions and postulates, and that is what many cosmologists are actually doing while you are reading this chapter, but all those models are necessarily over-simplistic and have little predictive power. In other words, they are able to describe in very general terms what is going on with existing observations, but they invariably fail and need to be readjusted whenever a new piece of observational data appears. So, in the end, cosmologists need to use numerical simulations and supercomputers to try to understand what is going on in the Universe.

My kitchen

What does it mean "to run a cosmological simulation?" I will try to let you peek into the kitchen of a numerical cosmologist, so that you can see what is being cooked and smell the aroma. Sorry, there is no way for you to taste a bit, unless you have extensive training at PhD level in computer science and astrophysics, have access to supercomputers, and are ready to spend years of your life doing very technical computer coding!

Our Universe is infinite, or at the very least extraordinarily large, so large that it can be considered infinite for any practical purpose. How can you hope to squeeze an infinite universe into a relatively small computer? Of course, there is no way. As a numerical cosmologist, you have to limit yourself to only a piece of the Universe. How large should the piece be? Well, if you take a piece the size of your room, it is not going to be very useful for cosmology, is it? In order to describe, or more precisely, model (i.e., describe quantitatively, on the basis of the known physical laws) how the Universe works and evolves, you need to model all the garden variety of various galaxies and the IGM as well, so you need to make sure that your piece of the Universe is large enough to include a few of each kind of the beasts. In other words, you need to make a Noah's Ark of the Universe. A simple boat will not work.

So you have made the ark. Let's now look inside it. What do you see? Well, just as a human eye can only see a certain level of detail in any picture, the same applies to a computer. Any computer model has a *finite resolution*. Let's imagine you are looking at the Moon. You do not see every last dust particle there, you can see only general contours of "seas" and "continents" and other details that are about 3,000 times smaller than the distance to the Moon. It turns out that modern cosmological simulations have "eyes" that are roughly three times as keen as a human eye; they can see details inside the Ark of the Universe that are about 10,000 times smaller than the size of the Ark. This number, 10,000, is usually called a *dynamic range*. And what matters for a numerical cosmologist is that this number is much too small.

In cosmology, we measure distances (and thus sizes) in very special units called "megaparsecs." This is a long word, and it means a very very large distance, more than 3 million light-years. If I were to write this distance measured in miles, my number would have 20 digits! Why is this number so large? Because so is the Universe! If you want to build a Noah's Ark of the Universe, the Ark needs to be at least 100 megaparsecs, or perhaps even 1,000 megaparsecs, in size. In comparison, the size of a normal big galaxy is only 1% of 1 megaparsec. Now we are ready to do the mathematics: if we want a 100 megaparsec Ark, we cannot resolve galaxies in it! It is precisely this factor of 10,000 to get from 1% (0.01) to 100. In such an Ark, galaxies will be at the very limit of computers' "vision." They will appear as bright points, but we will not be able to see their beautiful structure, elegant spiral waves, graceful thin disks, small but bright central bulges. Any detail smaller than the size of a typical galaxy will be lost to the computer's "eyes" in the same way that we cannot see individual mountains on the Moon. To resolve such details, we would need our dynamic range to be larger by another factor of 100: not 10,000 but 1 million.

Is this 1 million beyond our grasp? No! Do we need better supercomputers? Again no! The reason we are presently stuck with this dynamic range of 10,000 is not because our computers are not powerful enough (they actually are), but because we do not quite know how to do things right; we do not have good numerical methods and algorithms which can give us 1 million instead of 10,000.[5]

Having just told you this, I must admit that I was lying; or, more precisely, I was slightly out-of-date. What I have said was right yesterday, but it is wrong today. Science does not stand still: it always moves forward, and what was not possible yesterday is possible today; what is not possible today will be possible tomorrow. As you are reading this chapter, several groups of cosmologists are now working very hard on developing new numerical methods that will be capable of reaching this magic 1 million even on existing supercomputers. The name of the promising new technique is adaptive mesh refinement (AMR). It is a very clever way to improve the "vision" of cosmological simulations by another factor of 100, making it up to the desired 1 million (which will be much keener than human vision, or even the vision of a bald eagle). To be more specific, Figure 4.1 depicts in a symbolic form existing numerical methods for modeling the dynamics of cosmic gas in cosmological numerical simulations. As you can see, the variety of techniques is very large, ranging from simple Eulerian codes that keep all quantities describing the Universe on a uniform grid of cubic cells, to more advanced techniques such as smooth particle hydrodynamics (SPH) and arbitrary Lagrangian–Eulerian (ALE) schemes. However, the way to go is the future AMR, shown in the bottom right corner.

Why? The difference between AMR and other techniques is somewhat similar to the difference between a plane and a hot air balloon. Both carry you over the ground, but only the plane allows you to go wherever you like; with the balloon you are at the mercy of the wind. In a similar way, AMR is the only technique that allows you to control the flow of your simulation. With others you have to follow where the simulation leads you, but AMR gives you the freedom to put your computational

---

[5] This is not surprising if one recalls that numerical cosmology is only about twenty years old, and the first star was "formed" in a cosmological simulation a mere six years ago.

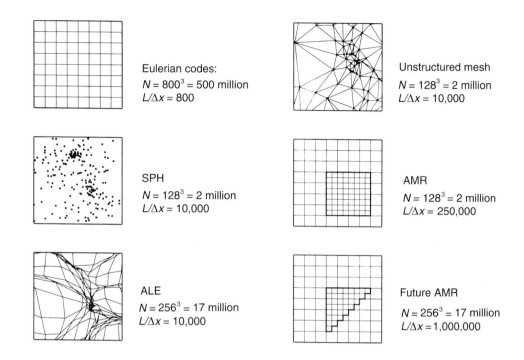

Eulerian codes:
$N = 800^3 = 500$ million
$L/\Delta x = 800$

Unstructured mesh
$N = 128^3 = 2$ million
$L/\Delta x = 10,000$

SPH
$N = 128^3 = 2$ million
$L/\Delta x = 10,000$

AMR
$N = 128^3 = 2$ million
$L/\Delta x = 250,000$

ALE
$N = 256^3 = 17$ million
$L/\Delta x = 10,000$

Future AMR
$N = 256^3 = 17$ million
$L/\Delta x = 1,000,000$

Figure 4.1   A symbolic representation of currently existing methods of cosmological gas dynamics. N is a characteristic number of resolution elements (cells or particles), and $L/\Delta x$ is the dynamic range achievable on the largest existing supercomputers. SPH smooth particle hydrodynamics; AMR adaptive mesh refinement; ALE arbitrary Lagrangian–Eulerian.

resources, your resolution, where they are needed most. By carefully managing your available resources, you achieve better results at the end, which of course is not surprising at all.

Let us now go back to the present. We do not quite have this desirable 1 million in dynamic range yet, we have to live with our 10,000 for a while. What can we do with this in the meantime? Just sit and wait until we finally get 1 million? Well, if we sit and wait, nothing will get done, that's for sure. Science does not jump from one place onto another, it develops at a steady rate, and we have to work with what we have now in order to progress to the future. Is there any use for simulations with a dynamic range of 10,000? Yes, a lot!

Managing the available resources

Current wisdom tells us that the bulk of matter in the Universe is hidden in an invisible form, the so-called "dark matter" (Jerry Ostriker in his chapter gives an exquisite account of its discovery). Frankly speaking, cosmologists do not quite know what this dark matter is made of, but they have at least some thoughts on the subject, and they know for sure that it is there. Since it is dark, it emits no light: that is what "dark" means. If it does not emit light, what does it do?

The majority of cosmologists think that this dark matter—whatever it is made of, black holes, parcels of dense gas, or undiscovered yet elementary particles—just sits around and produces gravity, and that's it. If this is all it does, it is rather easy to model! You still need supercomputers, but at least you do not have to worry about all the mess that normal ("baryonic") matter brings with it: shock waves, radiation physics, chemistry, star formation, nuclear physics, magnetic fields, etc. More than that, since galaxies are the sites where the light is emitted, all this baryonic mess is very important inside the galaxies, and cannot be ignored, but it can be ignored on larger scales, where the only physical process that takes place is gravitational attraction. Therefore we simply do not need this 1 million in dynamic range to model dark matter, we can simply use the 10,000 we have now. And this has been done.

This type of simulation, which models only the dark matter in the Universe, usually called N-body simulations, because they include a number (N, which can be very large, up to a billion) of "bodies" (black holes, gas blobs, stellar remnants, elementary particles, all the same!) which only interact gravitationally. Many N-body simulations have been performed to date (one of the best is shown in Figure 4.2) and one can say with sufficient confidence that the problem of the dynamics (but not the origin!) of dark matter is solved. In other words, right now, with existing supercomputers and numerical methods, we can run as big an N-body simulation as we would ever want (the biggest ones reaching a dynamic range in excess of 30,000, ten times better than a human eye).

Such simulations give scientists the whole picture of how structures form in the Universe, how various levels of structure—galaxy groups, clusters, superclusters—grow and interweave into a "cosmic web."

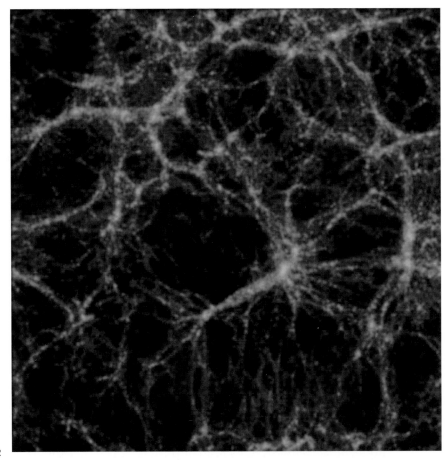

Figure 4.2

The dark matter density from a very high-resolution *N*-body simulation. Picture courtesy of J. Colberg (Max-Planck Institute for Astrophysics) and the Virgo Supercomputing Consortium.

Unfortunately (or fortunately for those who make a living by doing science), the Universe is not only dark matter. What we observe through our telescopes is light, and light is emitted by stars and other astronomical objects, which are made out of protons and neutrons, that is from baryons (a baryon is either a proton, or a neutron, or one of a bunch of very exotic particles found only in a laboratory). So what is next?

In order to model galaxies, how they form and evolve, we need to model what the gas filling up the Universe is doing, how it condenses into objects and fragments into small clouds, out of which the stars are born. For a numerical simulation this is not a simple modification: together with gas dynamics, we have to include a large variety of various physical processes: ionization and recombination of cosmic gas, radiative cooling and heating, molecular chemistry, evolution of the radiation field, etc. And of course, if we pretend to model galaxies, we also need to model star formation. The main problem with this is that we actually have very little idea how stars form in reality. In trying to include star formation in cosmological simulations, we have to resort to what is often called "sub-cell" modeling. It means we need to introduce some simple description of star formation based on the observational data and incorporate it in our simulations.

"What's the difference?" you may ask. Instead of using a simplistic model of the Universe, we resort to large supercomputer simulations, but at the bottom of these simulations is still a simplistic model, in this case the model of star formation. The difference is very large indeed. The variety of different galaxies is enormous, and the Universe looks very different in different places, whereas star formation is essentially the same everywhere (with perhaps only a couple of exceptions, which are easy to take account of). So in the long run it is much much easier to design a simplistic sub-cell model of star formation that will actually work than to design a simplistic model of the Universe. And that is what makes cosmologists believe that one day we will have a full and correct model of the Universe inside a computer.

But this time has not come yet. So what can we do with our current 10,000 of dynamic range? Quite a lot, even when complexities of the gas dynamics, star formation, radiation physics and chemistry are folded in. Yes, in the Noah's Ark of the Universe, we cannot resolve galaxies—they will appear in our simulations as points, dots of light—but we can study their distribution, how they cluster together to form groups, clusters, and superclusters. We can study how these structures change with time, and infer from these studies how and where galaxies form and evolve. Figure 4.3 (color section) compares the distribution of galaxies (bright points) from such a simulation (on the right) to the distribution of the dark matter (on the left). Those two distributions are different, and this difference is called "bias." Simulations like the one showed in the figure allow cosmologists to understand the main features of the large-scale distribution of galaxies and to relate the observed large-scale structures to the fluctuations in the cosmic microwave background and the physics of the early Universe which produced those fluctuations, thus moving a little bit forward in our understanding of the nature of the Universe we are living in.

But perhaps a more interesting way to use the current 10,000 of dynamic range is to look at what happens *inside* the galaxies. Of course, we cannot do this if we want to have the whole Ark, but we can do it for a small boat. In this way, we cannot hope to model all the variety of galaxies on all scales, but we can get a good peek at a few galaxies in a small region of space, or we can study the very beginning of galaxy formation, when there were only a few small galaxies, and then we can hope to manage with 10,000 in dynamic range instead of 1 million. And then unlimited possibilities open up for us.

Just imagine having the whole (albeit small) Universe in your computer! For this little universe you are the God, and it will play whatever tune you order it to. Want to change a law of physics? No problem; after all you are the God. Are you fascinated with the possibilities? I am.

And it actually does pay to change a law of physics. Let's imagine we want to understand what role supernovae—the end-of-life explosions of very massive stars—may play in the fate of a small galaxy. A galaxy in the sky is a galaxy in the sky, you cannot change or influence it, but a galaxy in your computer is in your complete power. Figure 4.4 (color section) shows an example of such an experiment. The column on the right shows the "right" universe, the one that is designed to model the real one. The left column shows a universe that has no supernovae: by an act of God (in this case, me), supernova explosions are forbidden. In the realistic Universe, supernovae blow up the whole galaxy, sending its gas and some of the stars into an empty space, but in the lame, supernova-less Universe, life is safe but dull.[6]

Simulations similar to the one described above can give cosmologists valuable insights into what physical processes are important at various stages in the evolution of the Universe, and which are not relevant. Such

---

[6] A movie of this simulation can be found at the following URL site on the web: http://casa.colorado.edu/~gnedin/GALLERY/sne_a.html

knowledge is essential if we are ever to have a complete model of the Universe inside a computer.

And, as is often the case when we infringe on an unexplored path, discoveries await us. Simulations are no exception, and what we sometimes cannot discover in the sky, we can discover inside a computer. The story of the heavy elements in the IGM is just one of many examples.

Making travel plans

The heavy elements (which for an astronomer means any element heavier than helium) can only be produced in stars and, before the first star was born, the Universe contained almost entirely pure hydrogen and helium. But after stars began to shine, after the first galaxies formed, the heavy elements—carbon, oxygen, nitrogen, and all the rest—were quickly dispersed within the galaxies, mixing into the primeval, pristine gas. Astronomers knew that for many decades, and no mysteries existed there. However, it was only recently, with the advent of super-powerful Keck 8-m telescopes at Mauna Kea in Hawaii, that new data started pouring in. And, to the great amazement of astronomers, heavy elements were detected observationally in the low-density gas far away from galaxies, in the Intergalactic Medium. How did they get there? Theorists did not wait for an answer: supernovae, in a gigantic explosion like the one depicted in Figure 4.4, can throw away the galactic gas, and all the heavy elements mixed in it, millions of light-years away. The problem seemed to be solved.

At about the same time Jerry Ostriker and I were working together on the early history of star formation in the Universe. We performed several numerical simulations and, in order to isolate the key physical effects, we forbade supernova explosions in some of our simulations. Can you imagine our amazement when we found that the heavy elements nevertheless found their way into the IGM with surprising ease? It took us a while before we realized that we had discovered a new way of transporting the heavy elements from inside galaxies to outside galaxies and putting them into the IGM. The physical mechanism is very simple: when two galaxies collide, pushed toward each other by their mutual gravitational attraction, they often hit each other with such force that chips of their gas get thrown away into the IGM, much the same way as an axe, cutting into wood, splinters wood chips all over the place. But in our case the axe was another galaxy, and chips were millions of times more massive than our Sun. Figure 4.5 (color section) illustrates this process: three panels show the collision of two small galaxies, as they approach each other (the left panel), hit (the middle panel), and the smaller galaxy passes through the larger one (the right panel), chipping a piece from it, which is pointed to by an arrow.

So, every time the heavy elements want to move into the IGM, they have a choice now: they can be thrown out by a supernova explosion, or chipped away by a collision with another galaxy. The choice, even if not too pleasant, is still better than no choice! And a year later, in follow-up work, which also included the simulation shown in Figure 4.4, I was able to demonstrate that the collision path is actually more efficient in delivering heavy elements into the IGM than supernova explosions. It is always nice to discover something new, but it is twice as nice to discover something new and important!

The list of examples is indeed very large. In the few last years, cosmological simulations led to a real breakthrough in our understanding of the formation of first galaxies and the evolution of the Intergalactic Medium between those first galaxies. Now cosmologists cannot wait to extend these simulations to more recent times to see how the majority of normal galaxies form in the Universe, and how those galaxies cluster together to create the unprecedented beauty of the cosmic structure.

The road ahead

So where do I stand now? Three years after I first crossed the threshold of Peyton Hall, my Alma Mater, I reclimbed the hill that I had descended, going from a graduate student to a postdoctoral researcher at MIT (still spending half my time at Princeton for two more years). I then took another postdoctoral position, at Berkeley, before becoming a faculty member at the University of Colorado. During all these years, I have run many simulations and, if nothing else, I have produced quite a few beautiful pictures and movies.[7] But the way forward is clear, and with the help of AMR and better physical understanding of what is going on in the Universe around us, cosmologists will have the whole Universe inside their computers within the next couple of decades. Not so much to become Gods, but rather to understand the world that we are a part of.

Suggested reading

Nick Gnedin's simulation gallery:
http://casa.colorado.edu/~gnedin/gallery.html
A popular article by Mike Schneider about Nick Gnedin's work:
http://access.ncsa.uiuc.edu/CoverStories/StarLight/starhome.html
The remarkable "Cosmos in a Computer" exhibit from the National Center for Supercomputer Applications (NCSA):
http://www.ncsa.uiuc.edu/Cyberia/Cosmos/CosmosCompHome.html
Matthias Steinmetz's gallery of cosmological simulations:
http://saguaro.as.arizona.edu/~matthias.html
Grand Challenge Cosmology Consortium home page:
http://zeus.ncsa.uiuc.edu:8080/GC3_Home_Page.html

[7] Isn't it great to direct a movie where galaxies and stars are the actors?!

PART II
Denizens of the Deep

5

The search for very massive black holes

DOUGLAS RICHSTONE, University of Michigan

 Ask an astronomer to name a theorist who observes, or vice versa, and Doug Richstone's name is sure to come up. Doug's first flirtations with astronomy resulted from a childhood fascination with the colors of stars in Orion. Despite a bicoastal education at Caltech and Princeton, he flourishes in the midwest as Professor of Astronomy at the University of Michigan. Doug is fond of saying that a busy research and teaching schedule, and too many committee trips, leave him little time for reading, hiking, and recreational travel. Despite this, he has accomplished something wonderful for this book: a fascinating essay describing the slow but nevertheless dramatic revolution in thinking about massive black holes and their role in the evolution of galaxies. In this essay Doug combines two of his career-long fascinations—the dynamics of stars and the nature of quasars—with his enjoyment of team play, to explore the black holes that lie at the center of so many galaxies.

When Alan Dressler called me in 1984, massive black holes were not on my agenda. I had known Alan since the mid-seventies when we were postdoctoral fellows, he at the Carnegie Observatories, I at Caltech. Although we hadn't worked together, his thesis, which included great observational work on clusters of galaxies, was very germane to the theoretical work I had done in my thesis, so I thought he chose good problems and did them well. More than that, from the weekly graduate students' basketball game, I knew he was someone you really wanted on *your* side. He got right to the point, "what's the answer?" The reply was too easy. "What's the question?" Now he was testy! It turned out that a week earlier he had sent a letter containing wonderful spectra of the centers of the nearby galaxy M31—the Great Nebula in Andromeda (visible on a summer's night as a fuzzy patch high overhead) and its companion M32. The letter had arrived that day and sat unopened in my mailbox. The data it contained were remarkable in two ways. It was the first time I'd seen a galaxy spectrum of the infrared calcium triplet, a beautiful set of three isolated narrow lines in an otherwise clean part of the spectrum, and the signal-to-noise of the spectra was at that time unprecedented, reflecting the emerging use of charge-coupled devices in astronomy. Alan wanted to know whether the rapid rotation of the centers of both galaxies indicated by the spectra implied the presence of a massive black hole in either. It was immediately clear that the data were superb and might yield an unambiguous answer, and that I was indeed in a good position to do the necessary calculations, so we plunged in. The calculations were difficult, and we were both doing other things, so the question took three years to answer.

Black holes in astrophysics

Although black holes are stunning consequences of the mathematics of Einstein's theory of general relativity, the basic idea actually predates general relativity, dating back to work by John Michell, Vicar of Thornhill, in 1784. In modern terms. we would say that Michell understood that in Newtonian gravity the escape velocity from the surface of an airless world (or star) is twice the square root of the mass of the world divided by its radius, times a universal physical constant, and is independent of the smaller mass being ejected. For sufficiently large masses or small radii he reasoned that the escape velocity could exceed the velocity of light, hence the star would be dark.

The subject lay fallow for two centuries, probably because Newton's corpuscular theory of light lost out to the wave theory, and because the fundamental role of the speed of light in dynamics was not understood until the twentieth century. Although black holes are implicit in the earliest solutions of the field equations of general relativity, it wasn't until the discovery of sources of high-energy radiation (the quasars and X-ray sources) in the 1960s that astronomers seriously began to consider the possibility that they were observing black holes. We now know that that the quasars are powered by very high-mass black holes (of more than one million solar masses), while some X-ray sources are powered by black holes (of a few solar masses) which are the remnants of now dead stars.

Quasi-stellar objects, or quasars, are smaller than the Solar System (hence their "quasi-stellar" appearance), but outshine the ten billion stars in a typical bright galaxy. They are the brightest steady-state luminous objects in the Universe. A significant fraction of this power is emitted in a well-collimated beam of high-energy particles. Within months of their discovery in 1963 the possibility that black holes might power these objects by accreting mass was suggested independently by Salpeter and Zeldovich, and these ideas have been studied extensively since then.

The attraction of black holes as an energy source for quasars is exactly that: their attraction. As a consequence of the conservation of energy, which underlies the escape velocity discussion above, an object dropped into a black hole will have enough energy to reach the speed of light as it vanishes. If some of that kinetic energy can be liberated as heat or radiative emission, a tiny object can emit an enormous amount of energy. Where most natural fusion reactors (stars) and man-made ones (bombs) achieve efficiencies of about 1% in conversion of fuel to energy (the Sun liberates about 0.7% of mc^2 for every ton of hydrogen it converts to helium), energy extraction by black holes can reach 10% and possibly more. Since quasars radiate about 10^8 solar masses worth of energy over their probable lifetime, every bit of that 10% is required. So, by 1984, when Alan called, most astronomers felt that the most promising model for these objects involved the heating of matter gravitationally drawn to a massive black hole. The production of a strong magnetic field in an accretion disk also seemed possible and the interaction of charged particles with such a magnetic field seemed to offer a plausible pathway to all of the observed phenomena.

The model had an important consistency check—black holes left behind at the death of the quasar should be observable in at least a few nearby galaxies. Since observations showed that the Universe was once heavily populated with quasars and these brilliant quasars have mostly died out, there should be lots of relic black holes present today. Even in 1984, a few nearby quasars were known to inhabit the nuclei of galaxies (the evidence is much stronger now; see Figure 5.1, color section) and the only way to eject a massive black hole would be to collect three or more of them in the galactic core and eject one by a three-body gravitational interaction. So the black hole accretion model of quasars makes a clear prediction that some fraction of present-day galaxies should have a massive black hole at their center. The idea of looking for these relics was first suggested in 1978 by Donald Lynden-Bell at Cambridge, and first attempted in 1979 by Wallace Sargent and Peter Young at Caltech.

Weighing invisible matter

The basic principle that underlies all searches for invisible matter is that ordinary matter still responds to it gravitationally. Any particle orbiting a putative black hole, or any other object, will travel at an average speed given by the formula below, and the mass (M) of the object (a planet, a star, or a black hole) can be determined by measuring a satellite's velocity (v) and the size of its orbit (R), according to the relation:

$$M = \alpha \left(\frac{v^2 R}{G} \right),$$

where G is the universal gravitational constant and where the parameter α has a value usually near 1. The value of α depends on the *distribution* of mass enclosed by the orbit, and computing it can be tricky. In the case of a star or gas cloud orbiting a black hole, we can often determine its velocity by looking at the Doppler shift of the spectral lines in the star as it moves toward and away from us. The Doppler shift is a change in frequency or wavelength of light waves (or sound waves) seen by an observer moving toward or away from the source of the waves (the pitch of a railroad whistle seems higher when the train is approaching and lower when it is receding). In general, then, increasing velocities near the center of a galaxy (bigger v's at smaller R), coupled with the absence of luminosity to account for that mass, have to be ascribed to an invisible object or objects.

In the particular case of M31, Dressler's observations revealed the

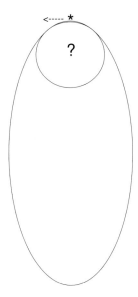

Figure 5.2

Two orbits of a star around a hypothetical black hole (at the question mark). The velocity of the star will be much greater in the plunging orbit because it has fallen toward the black hole. This effect is much larger if the black hole is embedded in a cluster of stars near the center of a galaxy.

average motions of a few million stars near the center of M31 and M32. The spectral lines from the galaxy clearly showed enormous velocities. The pattern of these velocities, however, was complex, indicating that a typical volume of space near the galaxy center had stars with large velocities in all directions, but with a net rotation about the center. Had we been observing a thin disk of objects (like Saturn's rings), it would have been reasonable to assume the stars were in circular orbits (to preserve the disk) but since the nuclei of both galaxies are fairly round we had to try to understand the orbits of the stars in detail. The problem was the possibility that many of the stars near the centers of both galaxies followed highly elongated orbits, traveling far from the center for a long time and then falling to and through the center at much higher speeds than nearly circular orbits would travel. Our problem was to tell clearly whether the high speeds we detected near the center were due to plunging orbits, or whether there was additional unseen mass acting on the stellar orbits (see Figure 5.2 and Figure 5.3, color section).

Our basic approach was to use a method developed by Martin Schwarzschild of Princeton University to study dynamics of galaxies. For any assumed *mass* distribution, we compute, one by one, the motion of individual stars in any (and all) bound orbits, storing the positions and velocities of each. Because the stars do not individually disturb each other appreciably, feeling only the average force of all of the others, we can add any mixture of stellar orbits to match the observed *light* distribution of the galaxy near the center. We then compare the velocities of the mixture to our observed velocities. An examination of the set of cases with varying central masses leads to a range of acceptable models. Sir Isaac Newton could have done any piece of this problem in the seventeenth century, but the number of calculations needed for this reliable but brute-force approach exceeds the capability of the lifetime of even a brilliant human. Since the calculations are very repetitive, a computer does them quickly and accurately without grumbling! The program that took an hour on my desk in 1984 now runs in seconds on a more powerful, and cheaper, machine.

The clear result, for a somewhat restricted set of models, was that we required very compact masses of a few million suns in M32, and a few tens of million suns in M31. Without that amount of mass, it was impossible to generate the large velocities observed near the centers of the two galaxies. Although these masses were too small to have powered the more luminous quasars that had motivated our search, they were the first convincing cases of black holes too massive to come from the death of individual stars, and very close to the quasar mass range. Although Alan and I had only demonstrated the presence of a small dark object or aggregate with great mass in the center of the galaxy, we optimistically suggested that this was, in fact, a black hole. As we wrote the paper, we noticed that the black hole masses in these two galaxies seemed to scale proportionately to the mass of the bulge of the galaxy. It proved to be a prescient observation.

Quite independently, John Kormendy (then at Dominion Astrophysical Observatory in Vancouver) had achieved rather similar results for M31 (by assuming the importance of nearly circular orbits in the galaxy and without the detailed model) and National Science Foundation produced a press release describing both his and our results. They were covered as front page news by the *New York Times*. I can still remember that it was the Monday (in July 1987) after a US-flagged tanker hit a mine in the Persian Gulf. I spent much of the day answering phone calls from science writers and assuring them that there was no danger we would be sucked into these black holes at any time in the future. Despite the excitement surrounding this result I had misgivings. After all, we'd found what we'd been after; could we have fooled ourselves? There were some unexplained features in the data, and the analysis had been done with spherical models. My view of science now is that we sometimes progress through optimistic interpretation, and I don't worry so much about a result that might turn out to be wrong because of some new development (in fact, I think it's vital to publish those results). Back then, however, I agonized over the possibility of getting "caught" in an error.

The "Nuker" team forms and grows

At this point (in 1987) we had a great method for prospecting the central parts of galaxies—spectroscopy of the infrared calcium triplet and the orbit superposition program—but we had only done the nearest two major galaxies (except our own, which requires different techniques), and we needed superior resolution to investigate more objects.

John Kormendy and Sandra Faber at UC Santa Cruz and her former student Tod Lauer (now at the National Astronomy Observatory in Tucson) had already put together the solution to our problem. They had decided to invest a significant effort in the study of galaxy centers with the soon-to-fly Hubble Space Telescope and reckoned that, to get lots of time to do systematic studies, a big diverse group was required. They added Alan and me and Scott Tremaine (then at the Canadian Institute for Theoretical Astrophysics).

The team was a very diverse set of personalities. At that point I wasn't very comfortable with writing a paper with more than two authors, and wondered how the team effort was going to translate into work versus credit. John seemed to be a superficially grumpy conservative, while Tod was gung-ho and rather excitable and struck me as callow (both impressions were superficial: my admiration of John and Tod has grown enormously over the last decade from watching their contributions to this work). Sandy had previously managed a big team that had obtained major results on the space distribution and motions and structure of elliptical galaxies; she was superbly organized and seemed to come in with a firm grip already on what we could hope to do. John and Tod sparred frequently over the choice of objects to observe; Scott and I often, as theorists, wondered why we were there, and I just can't recall Alan's early contributions. Despite my lack of experience with team efforts, I had a lot of confidence in this group. I had worked with both Alan and Scott and they were each flexible and capable, and John's overwhelming virtue was unerring care with the observations; when he said it was right, you could bank it. A few years later Sandy needed a collective name for us in an email message, and grasping for a word that characterized our study of the nuclei of galaxies she christened us "Nukers." The name has stuck. The six of us were the core of a group that would ultimately expand to fifteen.

The particular virtue the Hubble Space Telescope brought to the problem was its high angular resolution—the ability to see fine detail at the centers of galaxies. It achieved this resolution partly through excellent optics, but also because of its location in orbit. Above the Earth's atmosphere it could see the details of galaxy centers unblurred by the rapidly changing refraction produced by the turbulence of the atmosphere. We thought at the time that the typical quasar black hole would have a mass of about 100 million solar masses. In a typical big galaxy such a black hole would dominate the mass in ordinary stars only within about 50 light-years of the center of the galaxy. Thus, in order to see the fast moving stars hurtling by the black hole, we had to observe galaxies with resolution superior to this.

With ground-based telescopes in the 1980s, that limited us to the targets we had already observed plus an additional few. The Space Telescope offered much better angular resolution, which would translate to comparable spatial resolution for a large sample to the sort of data we had in the local group. Moreover, this resolution would be achieved in every single observation, not the few lucky nights on the mountaintop when the winds and weather cooperate. By taking spectra which characterized the motions of stars or gas in the galaxy nuclei on these fine spatial scales we could extend the work we had done on M31 and M32 to many other galaxies.

While Sandy and Tod were quite busy preparing for the launch of the telescope, and even busier after it, I ignored the subject and worked on clusters of galaxies. John, however, had moved from Canada to the University of Hawaii in 1990. By working with great care to exploit the spatial resolution obtainable with their telescope on Mauna Kea, John was able to push the search much further than I had expected. In late 1990, John contacted me with another interesting candidate. In this case, he had studied a galaxy 30 million light-years away called NGC 3115.

Again the story was similar, the rotation of the center, together with the random motions of the stars there, were simply too great for the center of the galaxy to hold together unless there was a much stronger gravitational field than could be provided by the stars. It turned out that there was indeed a very massive black hole (more than a billion solar masses). This was by far the most massive single object known at the time, easily capable of producing the luminosity of the most extra-ordinary quasars. At this point it began to dawn on me that we were being enormously successful. We had studied three objects in detail and had found something in all three objects (and, in addition, there was the galactic center and two others that John had noted). Once again the press was interested in the result, and John did a beautiful job of explaining what we had done, but a particularly able *LA Times* writer tracked me down in Kyoto by calling a series of colleagues I was visiting on the trip. I can only imagine the conversation with the hotel staff when she finally tracked me down. My Japanese colleagues were intensely amused by the entire episode. In my case the real fun came a couple of days later via email congratulations from Moscow, where two colleagues had learned the result from an Izvestia story.

The Space Telescope era

The launch of the Hubble Space Telescope in April 1990 (Figure 5.4, color section) was eagerly awaited and celebrated by a large fraction of the international astronomical community. Sandy and Tod, on the

Wide Field–Planetary Camera (WFPC) team, were among the first to realize that the telescope could not be properly focused. For our particular research effort, as for many others, this was a tremendous disappointment. The optical flaw which defocused the images made it impossible to take advantage of the location of the telescope above the atmosphere for dynamical spectroscopy at high resolution. Fortunately, we were able to use the telescope to systematically image the centers of scores of nearby galaxies. These data were supplemented by material obtained by the WFPC team. While this work proved ultimately to be of great value, I couldn't get excited about pictures without any information on stellar dynamics, and largely ignored these activities. Scott and I contributed some simple theoretical models, which were perhaps of some use in analyzing the data, but the team effort in the early 1990s was largely spent characterizing and trying to interpret the appearance, rather than the dynamics, of the centers of these galaxies, and was carried largely by Tod and Sandy. It was a bit frustrating, as we were doing taxonomy rather than physics, and our arguments with each other, often over nomenclature, seemed rather removed from the problem we were actually trying to solve.

The images did lead to an interesting debate about the ability of images alone to demonstrate the presence of a massive black hole. Our results showed that the centers of all galaxies had "cuspy" light distributions—rather than leveling out at a constant value, the density of stars in these galaxies seemed to climb inexorably toward the center. These sorts of structures could be easily understood if the growth of a massive black hole had pulled in the visible stars, or even if the gravitational field of the black hole pulled passing stars nearer it as they fell through the center of the galaxy. It was a tantalizing clue to some of the team members, while others, especially John, took the view that you simply had no idea what the mass distribution at the galaxy center was by looking at starlight without any information on velocities. The question of what one learns from only the light profiles of the galaxies remains interesting to me.

After 1993 there were several developments that stimulated progress in the area. While John's continued effort in Hawaii had kept me somewhat in touch with the black hole search, it was the repair of the

Hubble Space Telescope (HST) that really pulled me back into the subject. The repair itself was amazing. NASA had not demonstrated an ability to fix such a complex spacecraft in space previously and, as the mission approached, the laundry list of repairs grew, including not just the optics but a solar panel, panel drive electronics, and gyroscopes as well. I was very pessimistic. To my surprise, the repair was not merely successful, it was spectacular. Essentially all of the problems were corrected. I recall seeing the first pictures with 1,000 others at a meeting of the American Astronomical Society.

One of the images (of mass ejection from a massive star) was so detailed that the audience gasped. It was clear that we would be able to make the critical measurements on a number of galaxies. Sandy and John had led the way competing for telescope time and we would start to work on the black holes. We weren't alone. Working with the newly repaired telescope, Ford, Harms and their collaborators resolved what they thought was a disk of gas rotating at the center of the large elliptical galaxy M87 and measured its rotation near the center. Their results showed that gas clouds could be used as mass tracers at the centers of galaxies and indicated a 3 billion solar mass black hole (very close to an upper limit Alan and I had achieved with ground-based observations in 1990). Their work generated a great deal of attention and persuaded a lot of massive-black-hole skeptics that the objects were really there.

More important, however, was a less heralded object observed with ground-based radio telescopes by M. Miyoshi and his collaborators. They showed that the galaxy NGC 4258, which contained a low-power quasar-like source, had a very rapidly rotating disk near its center that contained MASER (microwave amplification by stimulated emission of radiation—the microwave analog of a laser) emitters. The radio observations made with multiple telescopes by Miyoshi's group gave resolutions far better than HST could, isolating the behavior of the central light-year in the target galaxy. In addition to providing evidence of (what was now) a modest mass black hole of 10 million solar masses, their work established that the mass was contained in a small volume. The only way to avoid a massive black hole was to invoke some as yet unknown astronomical object. This was a critical result because one could now argue that the massive dark objects we were discovering in other nuclei were indeed black holes by the Holmesian dictum "When you have eliminated the impossible, whatever remains, however improbable, must be the truth."

At this point John and I were writing an article reviewing the progress of the subject. John produced a plot, which he had actually published in a conference proceedings earlier, showing a correlation between black hole mass and galaxy masses—heavier galaxies had heavier black holes (see Figure 5.5). Indeed, the correlation is much better with the disks of spiral galaxies subtracted away (so that one compares the round bulge of the spiral to the dynamically and morphologically similar part of elliptical galaxies). It has never been completely clear to me who figured this out, or how (despite the fact that I coauthored most of the relevant papers). The idea was certainly discussed in 1988 in the first paper Alan and I wrote, and John and I certainly discussed it in 1990. John certainly made the plot first. Whatever the appropriate credit, the review article put the figure out in public view to a very wide audience. It is often reproduced in articles and talks on the subject (and in competing telescope proposals, without attribution!).

Yet, even four years later the figure's true significance is not clear to me. There are several mechanisms that could couple the growth of a central massive black hole to the bulge of a galaxy. Accretion of a fixed fraction of mass shed by dying stars, or of mass that might produce stars, could play a role. Either the formation of a bar or the absence of "centrophobic" orbits due to the presence of the black hole could transfer a roughly constant fraction of the mass of the galaxy bulge to the central black hole. Or it may be the formation process itself that traps the most tightly bound material into the black hole, or subsequent mergers of galaxies that scatter a fraction of the nearby stars into plunging orbits. Certainly the available reservoir of material depends on the mass of the bulge of the host galaxy. Is one of these critical or do they all play a role? Is this correlation the Rosetta Stone of supermassive black hole formation, or is it a minor clue? Whatever the answer, a model of formation and evolution of galaxy centers and black holes will eventually have to account for it.

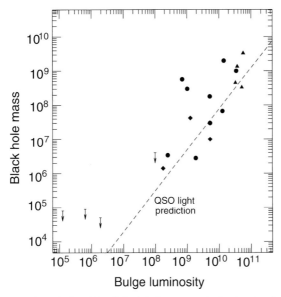

Figure 5.5

The relationship of black hole mass to bulge luminosity for all objects with dynamically measured central dark masses, as of October 1998. Mass and luminosity are expressed in terms of the solar mass and luminosity. Upper limits (i.e., non-detections) are shown as downward pointing arrows. The figure includes objects discovered by our team and by others.

The subject matures

Looking back over the last four years, I see that review article written with John Kormendy as a useful reference point. Since then we have demonstrated or greatly improved the case for very massive black holes in ten objects, and there are more than fifteen now with solid dynamical evidence. In three of the cases, including our own galaxy, the enclosed density of dark matter precludes aggregates of any known astrophysical objects. Thanks mostly to Tod, a database of more than 100 galaxies has exquisitely determined central light distributions, and they constitute an important resource for the field as well as possible targets for future dynamical studies. Karl Gebhardt, one of the newer "Nukers," showed by beautiful deprojection technique that there really seem to be two distinct kinds of elliptical galaxies in our sample (ones with very sharp cusps in the center (in the purely stellar distributions) and ones with very mild cusps, confirming a result that Tod and Sandy had achieved more easily (but less persuasively).

One important nagging problem has been the possibility that the relationship between black hole mass and bulge mass was a result of observational selection. To detect a small black hole, the host galaxy has to be fairly nearby so that we can probe it down to a small radius. On the other hand, none of the large bulge galaxies with big black holes was very nearby. This accident makes it possible for us to go wrong by making a series of small errors in interpretation of the velocity and then multiplying that by a small distance in the nearby galaxies (to get a small black hole mass) and a larger distance in the more distant ones. We devised a way to test the hypothesis that all galaxies contained massive black holes according to the relationship of Figure 5.5. The basic idea behind it can be thought of in the following simple way: if a galaxy has a certain probability of containing a black hole with a mass specified by a hypothesized model, then a particular galaxy, observed with a particular technique, has a detection probability (given the model) for a massive black hole.

We can assess the success of this model by looking to see whether we actually detect the black holes where expected. John Magorrian of the Canadian Institute for Theoretical Astrophysics, at the time the newest "Nuker," led us through an analysis based on this approach using heterogeneous data on about 30 objects. John also assumed that the centers of the galaxies were not dominated by plunging stellar orbits. The results were much stronger than I had expected, supporting the reality of the correlation shown in Figure 5.5 and indicating that nearly all galaxies have massive black holes.

Connections

So we have gone, over the last decade, from a time when serious people could argue about whether very massive black holes existed at all, and when black holes were thought of at best as rare and unusual features of nuclei of abnormal galaxies, to a time when few dispute the presence of supermassive black holes in the hearts of many galaxies, and many of my colleagues teach students in their freshman courses that central supermassive black holes are a standard feature of galaxies.

As I write this essay, the questions have changed. There is still some prospect that some of the individual objects we have studied will turn out, with better data, to be wrong. Its also possible that the rather sweeping results, which come as much from my team's considerations of results from competitors as from our own results, may turn out to be an optimistic reading of fragmentary data. But it seems to me that the question is no longer "are there massive black holes in galaxy centers?", and is instead "what are the effects of this previously unknown common feature of galaxies on their formation and evolution?"

It is too soon to try to answer this question, but there are some tantalizing clues already in the data. The census of the modern black hole population indicates that the total mass in black holes in the present-day Universe is larger than that required to produce the energy quasars radiate. It seems plausible that we see only 10% or 20% of the quasars that burn in the early Universe. The numbers of massive black holes are considerably larger than that, suggesting that quasars only shine for a few million years. Our observations of quasars in the Universe at modest age may only show us the tip of the iceberg. The quasar surveyors have already established, to my satisfaction, that the formation of these black holes is early, either coeval with the formation of the first stars in galaxies, or possibly earlier. The presence of the massive black hole, and the energy released from matter falling into it, must play an important role in the formation and evolution of the centers of these galaxies.

The most popular model for galaxy formation includes a few generations of "hierarchical merging" in which protogalaxies collide with others of comparable size and stick, building more massive objects. We know, from number counts and direct observation, that many galaxies in the Universe have merged at least once since the quasar epoch. Most of these galaxies must have already contained a very massive black hole. These black holes will inexorably settle to the center of the merger product, and, if they can shed their orbital energy to the lighter stars in the galaxy nuclei, the black holes themselves will merge. The mergers of comparably massive black holes are the most powerful events possible in our Universe, briefly outshining their galaxies by 15 powers of 10 (they are much more powerful than even gamma-ray bursts)! The deep irony is that these mergers—the most powerful explosions in the

Universe—are dark in the electromagnetic spectrum, radiating essentially all of the energy in gravitational waves. As I write this chapter several review committees are considering whether NASA and ESA (the European Space Agency) will join forces to fly an array of laser Doppler-ranging spacecraft which could detect ripples from these merging black holes in our distant past. I hope we will be able to "hear" their passing echoes before I retire.

The gravitational interaction of these massive black holes and the lighter stars nearby is a fascinating problem, and possibly an important one. The poet Wyslava Szymborska says "In my dreams I paint like Vermeer van Delft." In my dreams I can see the cosmic choreography of black holes and lighter stars in the galactic nuclei of our past. Perhaps understanding that dance is the *next* good problem to work on.

Suggested reading

Mitchell Begelman and Martin J. Rees, *Gravity's Fatal Attraction: Black Holes in the Universe*, Scientific American Library, 1998 (246 pages)
Kip Thorne, *Black Holes and Time Warps: Einstein's Outrageous Legacy*, W.W. Norton and Company, 1995 (619 pages)

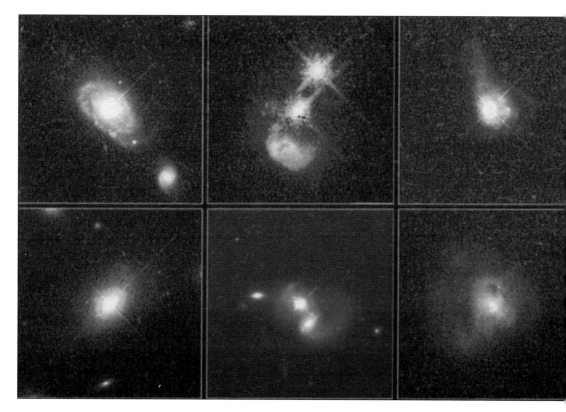

Figure 5.1. A collage of six low-redshift quasars imaged with NASA's Hubble Space Telescope by J. Bahcall (IAS) and M. Disney (Wales). These quasars are recognizable by the diffraction spikes indicating a point source, and are each in a galaxy.

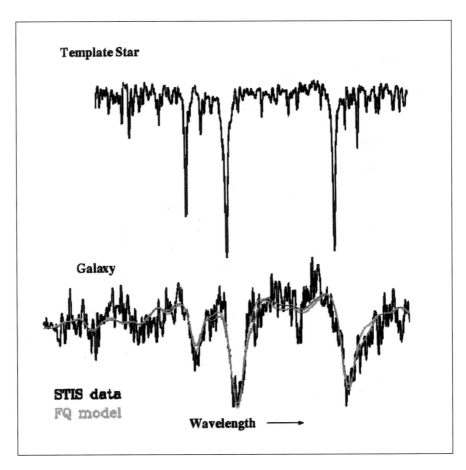

Figure 5.3. The spectrum of a galaxy and the type of star that dominates its light in the vicinity of the infrared calcium triplet. The three lines in the stellar spectrum are smeared and redshifted by the Doppler shifts of the million stars along this line of sight in the galaxy. These spectra were taken with the Hubble Space Telescope Imaging Spectrograph (STIS) by the STIS team. The model of the stellar velocities (the smooth curve labeled FQ) was made by fellow "Nuker" Ralf Bender of the Ludwig Maximilian University in Munich.

Figure 5.4. The aptly named space shuttle *Discovery* thunders aloft on April 24, 1990, carrying the Hubble Space Telescope to orbit.

Ground + X-ray

HST

Figure 7.1. (http://oposite.stsci.edu/pubinfo/pr/1998/26) ROSAT/Ground-based plus HST image of MS1054-0321. The X-rays (in blue) superimposed on a ground-based image of the monster cluster of galaxies, MS1054-0321, alongside a Hubble Space Telescope close-up of the galaxies in the cluster. This distant cluster is one of the most massive clusters in the Universe. MS1054-0321 and its sister clusters from the Extended Medium Sensitivity Survey provide strong evidence that the mass density of the Universe is insufficient to ever stop the Universe from expanding.

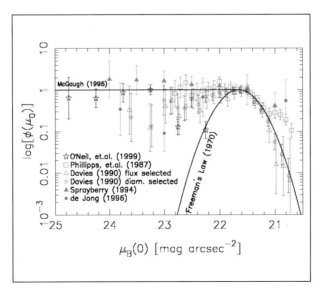

Figure 9.6. The space distribution of galaxies as a function of I_0. Various symbols indicate different surveys done by different observing teams. The solid black curve is the expectation of this distribution based on analyzing only high-contrast galaxies in catalogs. The basic result here is that the data suggest the space density as a function of surface brightness is flat: that is, there are just as many galaxies with $I_0 = 24$ as $I_0 = 22$. This means that these elusive galaxies are in fact quite populous in the Universe.

6

Gamma-ray bursts—the most spectacular fireworks

BOHDAN PACZYŃSKI, Princeton University

Bohdan Paczyński was raised and educated in Poland. He came to Princeton University's Astronomy Department in United States in 1982, where he holds the Lyman Spitzer professorship. Bohdan has held visiting positions in major astronomical institutions around the world, including Caltech, Cambridge, Harvard, Paris, and Moscow. This distinguished thinker and theoretician has been awarded numerous prizes for his contributions to astronomy, including the Gold Medal of the Royal Astronomical Society in London, the Henry Draper Medal of the US National Academy of Sciences, and the Eddington Medal of the Royal Astronomical Society. For over a decade now, he has concentrated much of his research in the study of one of the Universe's most challenging and enigmatic astrophysical phenomena: the bright, high-energy gamma-ray flashes known astro-colloquially as gamma-ray bursts.

Gamma-ray bursts (GRBs) were discovered with four US military spacecraft: *Vela* 5A, 5B, 6A, and 6B, launched in the late 1960s to monitor Soviet compliance with the nuclear test ban treaty. While first bursts were recorded in July of 1969, it took several years to develop proper software to uncover them from a huge volume of data, and the discovery paper by Ray W. Klebesadel, Ian B. Strong and Roy A. Olson of the Los Alamos Scientific Laboratory was published in *The Astrophysical Journal* on June 1, 1973. This became instant headline news for the astronomical community. By the time of the Texas Symposium in December of 1974 there were more theoretical attempts to explain them than there were gamma-ray bursts known at the time. Needless to say, these were mostly wild speculations, which is natural, as the observational results were so fantastic. The multitude of theories was a clear indication that nobody had a clue what was going on.

The bursts were very intense flashes of gamma-rays, some lasting several seconds, some up to several minutes, but no apparent pattern to their rapid variability, out-shining in gamma-ray domain the whole sky, including the Sun. Gamma-rays are photons, just like those that we see as optical light, but about one million times more energetic. Occasionally, a burst would be so intense as to disturb electronics in various spacecraft, or generate readily noticeable disturbances in the Earth's ionosphere, the outermost part of our atmosphere. If the same amount of energy was radiated as visible light then some bursts would appear brighter than any star, as bright as Jupiter or Venus. Six examples of GRB time variability are shown in Figure 6.1.

Perplexed

While theorists could not make sense out of the bursts, they were in fact predicted several years before their discovery, by Stirling Colgate, a maverick theoretician from Los Alamos National Laboratory. In an amazing case of a great insight or just sheer luck Colgate predicted in 1968 that supernovae explosions should generate short intense flashes of gamma-rays. Unfortunately, the details of his theory turned out to be incorrect, and the theory was dismissed. Today, some three decades later, Colgate's idea that GRBs are related to supernovae appears much more sound than any alternative, and in my view it is very likely to be correct.

In the late 1970s and throughout the 1980s, following several years of vigorous theoretical discussions, a long-lasting and incorrect consensus emerged: gamma-ray bursts were supposed to be caused by moderately energetic events on nearby, old neutron stars, at a typical distance of about 100 parsecs, that is about 300 light-years.

Neutron stars were discovered in the mid-1960s, and had been theoretically predicted in the mid-1930s. They are the end products of the evolution of massive stars, "ashes" left by some supernovae explosions. A neutron star has a mass 40% higher than solar, but its diameter is only 20 km; this is the size of a small asteroid. The huge mass and the small size imply a mind-boggling density, higher than the density within atomic nuclei, some 100 million tons per cubic centimeter. Some neutron stars are known as radio pulsars, some as X-ray pulsars, some as X-ray bursters. There must be millions of them in our Galaxy.

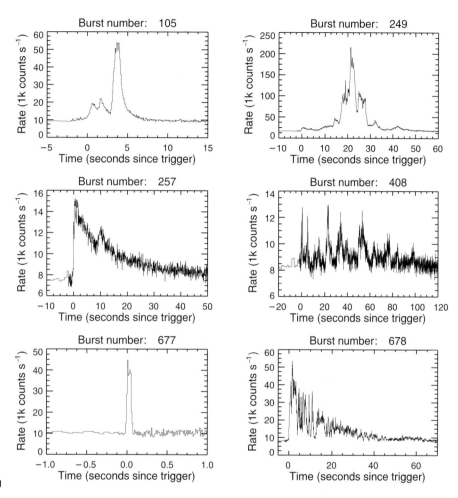

Figure 6.1

Six examples of diverse time variability of the intensity of gamma-ray emission of six bursts observed by the largest GRB experiment to date, BATSE, and kindly provided by Dr. Chip Megan of the BATSE team. Many more examples can be found on the World Wide Web at: http://www.batse.msfc.nasa.gov/

How were the neutron stars supposed to make gamma-ray bursts? Some similarity between GRBs and X-ray bursts provided a false clue. Discovered in the mid-1970s, X-ray bursts were one of the most spectacular success stories of modern astrophysics. Most of them were found in the sky close to the position of the galactic center, therefore their distance was estimated to be about 8 kiloparsec, or about 25,000 light-years. Their spectra peaked in the relatively soft X-ray domain, and looked like the so-called blackbody radiation with a temperature of 10 million degrees. Given the temperature, the distance, and the observed intensity it was straightforward to calculate that the sources had radii of 10 kilometers, just as expected of neutron stars. The energetics of X-ray bursts pointed to helium as the source of nuclear energy. It took S.E. Woosley and R.E. Taam less than two years to solve the puzzle and to explain X-ray bursts as nuclear flashes driven by helium "burning" just under the surface of neutron stars accreting matter from their companion stars.

Superficially GRBs had some similarity to X-ray bursts, with a roughly similar duration of several seconds or minutes. However, the GRB photons were much more energetic, their spectra clearly non-thermal, their time variability much more diverse, the observed intensity much higher; and GRBs never repeated, while X-ray bursts erupted every few hours or days. But it was argued that, if the GRB sources were much closer, at 100 parsecs rather than at 8 kiloparsecs, if the energy was released above the neutron star surface, rather than below it, if the eruptions repeated once every few decades rather then every few hours, if the nuclear fuel was accreted from the interstellar medium rather than from a companion star, then some analogy with the X-ray bursts might exist. Note, how many "ifs" appeared in the reasoning. Still, as far as I can see, this was the only argument that was at least somewhat rational. The models proposed to explain GRBs were purely hypothetical. No clear energy source was ever identified, no model has been calculated in sufficiently quantitative detail, and various proposed models were in direct conflict with each other. There was only one unifying concept: GRBs were believed to be associated with old neutron stars at a distance of about 100 parsecs.

Journey

For many years I did not follow GRB observations and theory in any detail, and from a distant perspective it all appeared reasonable. However, it all stopped looking reasonable at a closer inspection. I cannot recall the reason I got interested in the subject in 1986, and learnt what was known at the time about the distribution of GRBs in the sky. The distribution properties of any kind of sources were always of interest to astronomers, and they could be used to estimate the distance, even if the nature of sources was not understood at all. For example, when the X-ray bursts were found in the sky mostly in the direction of the galactic center, it was natural to assume that they were also the same distance away as the galactic center, at least approximately. If a supernova is observed in the direction of a nearby galaxy it is natural to assume that the supernova exploded in that galaxy. Of course, in some cases this type of association may be due to chance and we may be misled in case of any particular object or event. However, if we find clear statistical properties of the distribution of several hundred sources, then the chances for a mistake are reduced so much that the inferences based on the observed distribution are very robust.

Whenever astronomers opened up a new observational window using new instruments, a huge diversity of new types of sources was discovered in radio, infrared, ultraviolet, X-ray and gamma-ray domains. The first inferences about the distance to these sources were based on the observed distribution in the sky. If a particular type of object was found to be concentrated towards the galactic plane (i.e., in the Milky Way) it followed that a characteristic distance was several hundred or several thousand parsecs. If the objects were concentrated in the galactic center the implied distance was 8 kiloparsecs. If the sources were found to be distributed isotropically all over the sky, with no concentration towards nearby galaxies or clusters of galaxies, they were thought to be extragalactic, typically at a distance larger than 100 megaparsecs (i.e., more than 300 million light-years). Subsequent more accurate observations invariably confirmed these early distance estimates.

In 1986 it was well known that gamma-ray bursts were distributed isotropically and randomly all over the sky. This indicated that they were either very nearby, within about 100 parsecs, or very far, more than 100 megaparsecs away. In the former case the distance would be smaller than the thickness of the galactic disk, and we know that stars which are as close as that are distributed isotropically. In the latter case the distribution of all known objects is approximately isotropic as the distance scale is larger than that on which galaxies are known to cluster. However, any intermediate distance, between 100 parsecs and 100 megaparsecs, would show a significant departure from isotropic and random distribution. It was universally assumed that GRBs are nearby, even though there was no direct evidence to support that view. I wondered if there might be some way to figure out which of the two vastly different distance ranges was indicated by the observations.

Very important information was provided by the distribution of the apparent GRB intensities. There were few strong bursts, and many more weak bursts. In general, we expect weak sources of any kind to be farther away, and therefore detectable over a much larger volume. Hence they are more numerous than strong sources. This trend was very clear for strong and moderately weak bursts. However, the number of very weak bursts was surprisingly small, as if there were very few of them beyond some distance.

A modern representation of these two distribution properties is shown in Figure 6.2 and Figure 6.3. These are the results obtained with the instrument called BATSE (Burst And Transient Source Experiment), developed by Gerald J. Fishman and his group at the NASA Marshall Space Flight Center in Huntsville, Alabama. The results available in 1986 were based on fewer bursts, but appeared similar. The conclusion was obvious: as the sky distribution is isotropic, but only few very weak bursts are detected, we must be located at the center, or near the center, of a spherical distribution of sources. The "flattening" of the number counts in Figure 6.3 implies that at some distance we are reaching the limit beyond which there are very few, if any, gamma-ray bursts. The isotropic distribution implies we are near the center. These two figures exclude the galactic origin. All known kinds of stars, all interstellar dust and gas observed in our Galaxy is concentrated either to its center, or to its equator. This is not the case for gamma-ray bursts.

So, where are the bursts located? One may search all textbooks and all scientific papers about the distribution of all kinds of astronomical objects to find that there is only one possibility known to us: the sources must be at cosmological distances, far out of our Galaxy. Why? Because it is only the whole observable Universe that has the property that we appear to be located at its center. Why? Because we see the Universe expanding at the same rate in all directions. This expansion

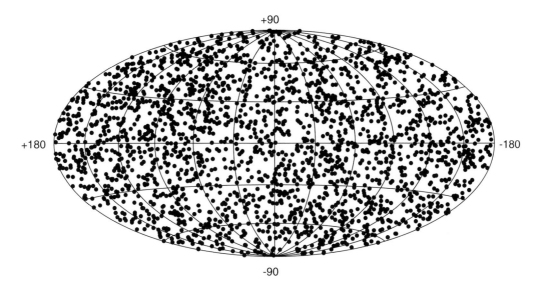

Figure 6.2 The observed distribution of 2365 gamma-ray bursts detected by BATSE in the first six years of its operation is shown in a galactic coordinates system; the galactic center is at the center of the figure. Note that the distribution is isotropic and random. This figure was obtained from the BATSE site on the World Wide Web at: http://www.batse.msfc.nasa.gov/

began some 14 billion years ago, and this imposes the distance limit: we cannot detect any source which is farther away than 14 billion light-years as there has not been enough time for the light emitted by such a source to reach us yet. If we would ever like to see a more distant Universe we have to wait. As the Universe gets older we can see farther and farther away. But at any time we can only observe a finite volume of the Universe, and we appear to be at the center of that volume.

This was a textbook reasoning, there was nothing original about it. Yet, for some reason, it was not accepted in the case of gamma-ray bursts. When I presented it at several conferences all I heard back were sarcastic remarks, and my scientific reputation was shaken a bit. It was not all serious. During a reception following one of the conferences I made a bet with Ed Fenimore of the Los Alamos Laboratory: he claimed the bursts were galactic, I claimed they were at large, cosmological distances. The loser was to provide the winner with a bottle of wine, though the quality of wine was not specified.

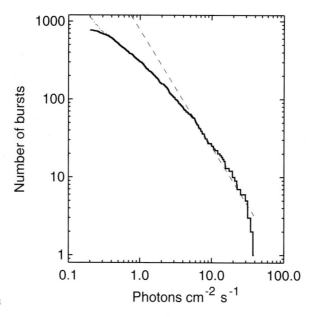

Figure 6.3

The distribution of the observed intensities of gamma-ray bursts detected by BATSE, as kindly provided by Dr. Chip Megan of the BATSE team. Note that, at the upper left corner of this figure, which corresponds to weak bursts, there are fewer than expected from a uniform distribution of sources in Euclidean space. This implies that we see evidence of the finite extent of the distribution.

Things did not look good for some time. First, I discovered that the equivalent of Figure 6.3, as known in 1986, was done incorrectly. When the proper quantity was used as a measure of the apparent burst intensity there was no evidence for the flattening of the counts. My reasoning was sound but it was based on an incorrect diagram, which of course made my conclusion unjustified. I also found that I was not the first to reach the conclusion that GRBs are at cosmological distances. Russian astrophysicists V.V. Usov and G.V. Chibisov had reached this conclusion in 1975, using the same reasoning, also based on an incorrect form of Figure 6.3. It was clear that the observations available in the late 1980s were not sensitive enough to reach the distance beyond which there were either few or no GRBs. We had to wait for a new, more sensitive instrument, which was under development by Gerald Fishman, Chip Megan and their team at NASA Marshall Space Flight Center.

Megan and Fishman tested their huge detector in a long balloon flight. With the very high sensitivity they expected to detect several dozen bursts, but they found only one. It was clear that their instrument was sufficiently sensitive to reach the limit of the GRB distribution. However, with a single detection there was no way of knowing if those very weak bursts were concentrated in the Milky Way, as expected if they were in the galactic disk, or all over the sky, as expected if they were at cosmological distances. It was necessary to wait for the new detectors to be launched as part of Compton Gamma-Ray Observatory (CGRO) into space, and to accumulate a large number of weak bursts.

Compton's wine

In the spring of 1991 the Compton Gamma-Ray Observatory was launched, with BATSE on board. In a few months enough bursts were detected to obtain two critical diagrams, like Figure 6.2 and Figure 6.3 in this chapter. The two diagrams, with the distribution of 143 BATSE bursts, were presented by Chip Megan at a conference in Annapolis, Maryland, on September 23, 1991. The correct units were used to measure burst intensity, so that there was no mistake in interpreting the results. It was clear that the bursts were not concentrated in the galactic plane or the galactic center, and there were few very weak bursts. This was a textbook case for the cosmological distance scale. It was by far the most exciting moment in my professional life—my guess was correct, GRBs were far away.

Approximately 50% of the participants at the Annapolis conference instantly recognized the implications of the BATSE results. Some were not happy at all, as they had worked for many years on the galactic models of the GRBs, but they realized there was no way to insist that the bursts were in our Galaxy.

It was amazing that the other half of the participants remained unconvinced. For several years they kept making various attempts to somehow keep the bursts within our Galaxy—if not in the galactic disk, some 100 parsecs away, then at least in the extended galactic halo, some 100 kiloparsecs away. The debate continued, conference after conference. Finally, in the spring of 1995, Robert Nemiroff organized a formal debate in Washington DC. I argued for the cosmological distance to the bursts, Don Lamb of the University of Chicago argued for the galactic origin. There was no decisive swing in the opinion of several hundred participants of the event.

A breakthrough came two years later, following the launch of the Italian–Dutch spacecraft named BeppoSAX. Gamma-ray bursts were not on its scientific program. Yet, within several months it became clear that the instrumentation was almost perfect for providing fairly accurate positions of GRBs fairly fast, and that turned out to be critically important. Before BeppoSAX it was possible to obtain positions from BATSE within several seconds of the beginning of a burst, but the location was known to no better than several degrees; that was too large an area to search for optical and radio counterparts. It was also possible to obtain positions accurate to better than an arc minute from other instruments, but only days or even weeks after the burst. Many searches were made for the optical and radio sources which might be associated with GRBs, but none succeeded. The error boxes were either too large or they came too late. And now BeppoSAX provided positions with errors of a few arc minutes only, and within just several hours of the burst. It turned out that this combination of accuracy and speed was essential, and X-ray, optical and radio counterparts were found to about two dozen gamma-ray bursts. These sources were initially fairly bright, but they faded rapidly, and they were named "afterglows."

The decisive event came on May 8, 1997, with the burst named GRB 970508. Its optical afterglow was discovered by Howard Bond with a 1-meter optical telescope at Kitt Peak National Observatory. A few hours later M.R. Metzger obtained the spectrum with the Keck 10-m telescope located on top of Mauna Kea in Hawaii, and established that the source had several spectral lines with their wavelengths increased by a factor of 1.835. In other words this afterglow had a cosmological redshift $z=0.835$, demonstrating clearly that the gamma-ray burst was indeed very far, several billion light-years, away from us.

Since then very accurate positions have been obtained for at least 16 optical or radio afterglows. In almost all cases a very faint galaxy was found at those locations as soon as the afterglow faded. Some of those galaxies were so faint that only a Keck or Hubble telescope could detect them. In two or three cases the so-called "host galaxy" was too faint even for these powerful instruments. Nine afterglows and/or their host galaxies had their redshifts accurately measured and found to be in the range 0.43 to 3.42, about as far as the distant quasars and galaxies. There is no longer any doubt about which distance scale is correct. At the recent conference in Santa Barbara I was given a bottle of wine by Ed Fenimore in recognition that I won the old bet. The wine came from Los Alamos, and it had the name: "La Bomba Grande."

I won the bottle of wine, but I missed an opportunity to win another, much more serious bet I was offered by Sir Martin Rees, the leading British theorist. At the beginning of September 1991, just two weeks before the presentation of BATSE results at the Annapolis meeting, he gave the odds 100:1 that the bursts were galactic. A few years later Sir Martin remarked: "we were both fools, I for offering the bet, Bohdan for not accepting it."

Of supernovae and hypernovae

This may not be the end of the story, as there are at least three different kinds of gamma-ray bursts: short, lasting less than 2 seconds, and long, which come in two types: hard and soft. Short GRBs, and also long and hard bursts, have most of their energy in very hard gamma-rays, while long and soft bursts have no high-energy photons. BeppoSAX detects only GRBs of the long and hard variety, and all afterglows discovered so far are related to this type. Also, this is the only type of GRBs for which there is a clear shortage of very weak bursts, as expected of any kind of sources detectable throughout the Universe. However, the other two kinds—short GRBs and long but soft GRBs—show no evidence of a deficit in the number of very weak bursts. Therefore, it is not certain that these also come from cosmological distances.

Sometimes theory works, at least in a broad sense. The afterglows of gamma-ray bursts were predicted by simple theoretical considerations. The principle is simple: whenever we observe a powerful explosion in the Universe, soon afterwards we detect the effects of collision between the matter ejected by the explosion and the gas which is filling all space. Such phenomena are observed following supernova explosions, and the large bubbles of hot interstellar gas they create are known as supernova remnants. Powerful explosions known as quasars generate jets of gas streaming close to the speed of light; when these collide with intergalactic gas then truly gigantic blobs of hot gas are formed, and they radiate very strongly in the radio domain. Therefore, it was natural to expect that gamma-ray bursts should be followed by an emission lasting for several weeks or even longer. Simple models were calculated by me and my student James Rhoads in 1993, and more detailed models were developed by other theoreticians in 1994 and 1996, predicting that there should be radio, optical and X-ray sources at the location of gamma-ray bursts, fading rapidly with time. It was wonderful to see the discovery of GRB afterglows in the spring of 1997.

While the afterglows are more or less understood, and we are learning a lot with a steady stream of new observations, the bursts remain enigmatic. There is no doubt that these are violent explosions with the debris ejected at nearly the speed of light, but we do not know what kind of star (if stars they are) is so explosive, and how is it possible that so much energy is radiated as gamma-rays? The two most popular models claim that these are either pairs of neutrons stars merging and exploding, or that these are extremely powerful explosions similar to supernovae: violent deaths of massive stars. While neither model can make quantitative predictions about gamma-ray emission they make reasonably firm predictions about the expected location of the bursts.

Pairs of neutron stars are well known to exist. They are the end products of the evolution of massive binary stars, which have consecutively exploded as supernovae. Theory predicts that a pair should move with a velocity of several hundred kilometers per second, and in fact the known pairs of neutron stars are observed to move that rapidly. According to the general theory of relativity any pair of masses orbiting each other radiates gravitational waves. This phenomenon has been already observed: R.A. Hulse and J.H. Taylor of Princeton University were awarded the Nobel Prize for demonstrating that a pair of neutron stars was losing energy at the rate predicted by the theory. Energy loss brings the two stars closer together, and ultimately they merge forming a black hole with a small amount of matter ejected explosively. But the gravitational radiation is very weak and it takes a long time for the pair to merge, typically between 100 million and a billion years. During this time the binary star travels a long distance from the place it was born, and dies in a typical cosmic environment, with very little gas or dust around it.

The alternative model, the violent death of a massive star, is expected within one or two million years of a star's birth from the condensation of a dense cloud of interstellar matter. Therefore, the explosion should take place close to where the star was born, in an environment full of interstellar gas and dust. This is distinctly different from what is expected of a neutron star merger.

What about observations? They are not definite yet, but seem to favor a "dirty" environment at the sites of GRB explosions, and provide tentative support for the death of a massive star as the correct model. Within several years, with a vastly larger number of well-studied bursts and their afterglows, we should find a definite answer to the question: are the bursts associated with "dirty" environments, characteristic of regions in which new stars are born, or are they preferentially in typical galactic environment, that is in space with little gas and dust? If the current indication is confirmed then we shall know that the original idea of Stirling Colgate was correct as a general concept, associating gamma-ray bursts with extra-powerful supernova explosions. The name for these hypothetical events is already popular: they are called hypernovae.

What is a hypernova? It is certainly a reasonable name for an explosion far more powerful than an ordinary supernova. It is also a testable astronomical concept: the most violent explosions are expected to be associated with deaths of the most massive stars. Such stars live for a short time only, and therefore they are expected to explode near the place they were born. It is far more difficult to establish the physical nature of such powerful explosions. Popular ideas include rapid stellar rotation, gravitational collapse and superstrong magnetic fields. While such theoretical models should be developed, and many theorists are working on such models, the past history of GRB theory is not very encouraging. There is some evidence that gamma-ray bursts are related to hypernovae, and in a few years we shall have firm observational confirmation (or refutation) of this idea, but I am rather skeptical that we shall be able to develop a convincing and sound model of these powerful explosions any time soon.

There is already one case of a long and soft burst that appears to have an optical counterpart: GRB 980425 was detected within several arc minutes of a very powerful supernova named SN 1998bw, which exploded almost simultaneously in a relatively nearby galaxy with the redshift $z = 0.008$, i.e. about hundred times closer to us than a typical GRB afterglow. While the burst was soft and much weaker than a typical long burst, the supernova was 20 or even 30 times more powerful than an ordinary supernova. This remarkable coincidence opens up a possibility that the more powerful GRBs detected at much larger distances are associated with even more powerful explosions, which might be called hypernovae. Indeed, there is evidence that those distant supernovae are associated with star-forming regions, where we expect the most massive stars to be born and to die violently. And so the old prediction by Stirling Colgate becomes relevant. He was wrong about the detailed model, but it appears that he might have been right in a broad sense: some, perhaps many, perhaps even all, gamma-ray bursts are associated with hypernovae explosions.

Let me explain the supernovae. There are two main types of supernovae: type Ia are nuclear explosions of massive old stars known as white dwarfs. All other types, including the most common type II, are caused by a violent release of energy following a formation of a hot neutron star at the end of the life of a massive star.

White dwarfs have a mass comparable to that of our Sun, but they are as small as Earth, which makes them very dense. A nuclear explosion is triggered when central density exceeds about 1,000 tons per cubic centimeter. The whole white dwarf is disrupted, its debris ejected with a velocity of about 10,000 kilometers per second, and a total kinetic energy of the order of 10^{51} ergs. This is about as much energy as our Sun radiates during ten billion years, that is, during its entire lifetime.

Type II supernovae are even more spectacular. When the inner core of a star 20 or 30 times as massive as our Sun runs out of nuclear fuel, there are no energy sources left, and the core collapses under its own gravity. The collapse is halted only when the stellar density exceeds the density of matter in atomic nuclei, i.e., about 100 million tons per cubic centimeter. Matter as dense as that is hard to compress, and a hot neutron star is formed. It has a mass about 40% higher than solar mass, but it is only 20 kilometers across. It is very hot, with a temperature of 100 billion (10^{11}) kelvin, as 10% of its mass has been converted into thermal energy released by its gravitational field. This huge energy of about 3×10^{53} ergs is mostly lost in just over 10 seconds in a powerful neutrino/anti-neutrino burst, and only a small fraction, about 10^{51} ergs, is somehow transferred out and gives rise to the explosion of the rest of the massive star.

It is useful to keep in mind that, while we have a good understanding of various types of supernova explosions, our knowledge is based on a patchwork of theory and observations. Theory alone cannot predict quantitative properties of any supernovae, as they are so complicated. Gamma-ray bursts are even more complicated, and while the volume and diversity of observations increases rapidly, it is no match for what we know observationally about ordinary supernovae. Therefore, it is not likely that the GRB puzzle will be solved by theoretical means alone. We shall need a lot of guidance from the most diverse and ingenious observations.

A spectacular case of a new type of possible observations was demonstrated on January 23, 1999. A very strong gamma-ray burst was recorded by BATSE—its time variability is shown in Figure 6.4. The on-line computer system automatically calculated the location of the burst in the sky in a matter of one or two seconds, but with an accuracy no better than about 10 degrees. This information was automatically emailed to another computer in Los Alamos, and a small instrument was pointed in the general area of the sky to take a series of short optical exposures. All the observations were recorded and archived automatically. Everything was pre-programmed, no human intervention interfered with the computers working on-line. The data would have stayed in the archive vault forever if it had not been for BeppoSAX, which detected the same burst, and several hours later provided coordinates accurate to several arc minutes. Many observers rushed to their telescopes and the optical afterglow was discovered. Given a good position, the data archived at Los Alamos were looked at and an amazingly bright, rapidly variable object was discovered. All the optical measurements are shown in Figure 6.5.

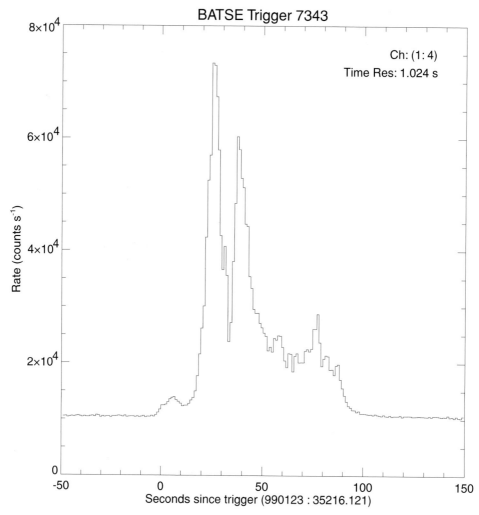

Figure 6.4   One of the most intense gamma-ray bursts detected by BATSE and other instruments on January 23, 1999. This figure was obtained from the BATSE site on the World Wide Web at: http://www.batse.msfc.nasa.gov/

What does it all mean? Spectra of the optical afterglow revealed a cosmological redshift of $z = 1.60$, that is a distance of about 10 billion light-years. Yet, in spite of this enormous distance, the GRB 990123 was one of the strongest bursts on record. The optical flash, recorded while the gamma-ray emission was still prominent, corresponded to a star of 9th magnitude. This could have been seen with ordinary binoculars. Just imagine: to see with binoculars a flash from such a distance! Had

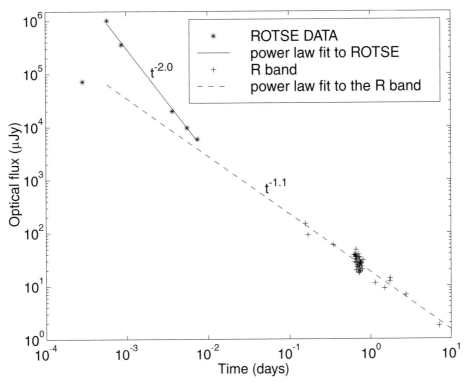

Figure 6.5   The optical flash detected by the ROTSE instrument at the time of the strong gamma-ray burst on January 23, 1999 (the first 0.01 day, i.e., the first 15 minutes), and the optical afterglow detected with many instruments after 0.12 day (about 3 hours), adapted from the paper astro-ph/9902009 by Reem Sari and Tsvi Piran. The vertical scale represents the optical intensity, with the top corresponding to magnitude 9 (detectable through binoculars), the bottom to magnitude 24 (detectable with large telescopes only). This GRB 990123 was at a distance corresponding to the redshift $z = 1.60$, that is, approximately 10 billion light-years away from us.

the burst been within our Galaxy, at a distance of 1 kiloparsec, that is about 3,000 light-years away, it would shine as bright as our Sun at high noon! And remember: there was 10,000 times as much energy in its gamma-ray radiation.

It is also amazing what instrument was used to make this discovery. The project called ROTSE (Robotic Optical Transient Search Experiment) is led by Carl Akerlof of the University of Michigan, and participants at Los Alamos National Laboratory and Lawrence Livermore National Laboratory. (Their Web site is: http://www.umich.edu/~rotse/) The instrument is made of four CCD (charge-coupled device) cameras, each using a Canon telephoto lens, just 11 cm diameter. This is a low-cost instrument, small by any astronomical standards. Yet, thanks to the ingenuity of the team, the automation, the clever software, it made the discovery of the year. It demonstrates that progress in astrophysics, including the difficult field of gamma-ray bursts, can be made with low-cost small instruments. I think it is only a matter of time before some Super-ROTSE makes an independent discovery of a distant GRB by detecting its optical flash. Who knows how many other unexpected discoveries there are in the future?

Suggested reading

Peter J. Leonard and Jerry T. Bonnell, Gamma-Ray Bursts of Doom, *Sky and Telescope*, February 1998, p. 28

Articles by Robert J. Nemiroff, Virginia Trimble, Gerald J. Fishman, D.Q. Lamb, Bohdan Paczyński, and Martin J. Rees, in *Publications of the Astronomical Society of the Pacific*, December 1995, Vol. 107, No. 718 (special GRB debate issue)

7

Clusters of galaxies and the fate of the Universe (Or how to be a cosmologist without really trying)

MEGAN DONAHUE, Space Telescope Science Institute

Megan Donahue makes her living studying clusters of galaxies and intergalactic gas, and tending the on-line data archives of the Hubble Space Telescope at the Space Telescope Science Institute. She lives in Towson, Maryland, with her astronomer husband Mark Voit, and their two children, Michaela and Sebastian. Megan was born and raised in rural Nebraska, was an undergraduate in physics at MIT, and earned her PhD in astronomy at the University of Colorado Boulder. She went on to postdoctoral positions at Carnegie Observatories and the Space Telescope Science Institute, where she works now as a staff astronomer. Megan is a bright light among young extragalactic observers, and the coauthor of the astronomy textbook, *The Cosmic Perspective* by Jeffrey Bennett, Megan Donahue, Nicholas Schneider, and Mark Voit (Addison-Wesley, 1999). Here, she tells us the intertwined story of her own coming of age in science, and a trail of clues that is leading us toward a better understanding of galaxy clusters.

The concept of Fate makes me nervous. Yet, with a handful of observations made from our tiny corner of the Galaxy, we can determine the fate of the entire Universe. We have known since the late 1920s that the Universe is expanding. But what we are just beginning to discover is whether the Universe will expand forever or will eventually stop expanding and collapse in on itself.

I have played a supporting role in the quest to reveal this fate—my observations of a handful of distant clusters provide some of the strongest evidence we have today that the Universe will expand forever. This is a startling statement about the Universe—not startling because it is unexpected, but startling in that we, sitting here on the planet Earth, can make a statement like this at all. With physics on my side, I have an easier time predicting the fate of the Universe than I have predicting the trajectory of my own life. My story will tell you what I mean.

A story about a scientist usually begins with a statement about how much a Boy of Destiny he was as a child. I didn't even want to be an astronomer when I was a kid. I wanted to be gymnast, like Olga Korbut. I also wanted to be a jockey. But my final growth spurt at the age of twelve rather settled the issue—I was going to have to change career plans. I began to practice my jump shot.

Growing up on a corn farm in Nebraska, I had very little exposure to science or engineering. I had never met a professional scientist before I went to college, although my best friend's father was an engineer. I certainly didn't want to be a farmer. The first time I began to consider physics was during that final scene in *Planet of the Apes*, when Commander Taylor rides up a beach on a horse and sees the Statue of Liberty buried to its waist in the sand in the aftermath of a nuclear apocalypse. I had been paying attention to the horse and the desert, and wondering if Taylor was going to have to eat his horse; the scene left me thinking I'd better learn about nuclear physics if the Bomb had the potential to rearrange our coastline. In the late 1960s and early 1970s I had heard a lot of grown-ups talk about the Bomb, but I hadn't appreciated its terrible potential until that moment.

So I changed my reading habits from horses to science fiction and physics. School science classes were too slow, too qualitative and, worst of all, too infrequent. So I began to try to teach myself relativity in sixth grade. But I wasn't some child genius—so I ran aground very quickly. On units. I was contemplating the equations $E = mc^2$ and $F = ma$. What does it mean, I pestered my sixth grade science teacher (a patient and sainted man named Mr. Simmering) to multiply pounds by 186,000 miles per second twice? What kind of crazy number was that? I looked up the English unit of mass and was rewarded with the answer slugs. What kind of system was this and what were slugs? He replied that physicists use the Metric System and the appropriate units were kilograms and meters or kilometers per second. I thought about that for a moment and replied, frustrated, "well, that's pretty arbitrary!" And the slugs thing left me fairly cold. Could physicists just make things up? That seemed unlikely, there had to be some secret to this. Maybe you did have to be an Einstein.

I had big hopes for junior high school. Junior high science could happen 3–5 times a week, plenty of time to be illuminated about the mysteries of science. Surely learning things in class would be easier than what I had been doing. And junior high science teachers—they had to know just about everything! I finally got the nerve to corner my teacher Mr. Alderman (in retrospect, a youth just out of college, but to me then a wise man replete with the knowledge of the Universe) with my burning question: What makes an electron have charge? I knew that the balance between the numbers of protons and electrons gave positive or negative charge to large objects, but what made an electron negatively charged in and of itself? He smiled, but after a lot of grim pressure on my part, he admitted he didn't know.

Teachers, I discovered, even my hallowed science teachers, didn't know everything! It was then that I began to appreciate that not all was known about the Universe, that there was more knowledge out there than could be easily gained by reading books and pestering teachers. That bothered me for a little while because I am a lazy learner at heart.

So began my quest to be a scientist. I didn't build a telescope, but I did configure a Morse code transmission and reception station and string up a dipole ham radio antenna along the length of our one-story ranch house. I didn't know that junior high students rarely did this; I read about amateur radio from a library book and decided that would be fun to try. My parents advised me to write a letter to the newspaper (I have no idea why they told me to do this). When I did, a man named Pat Patterson responded with a course for kids in amateur radio (I surprised him by being a girl), and I was on my way. I have been consistently lucky in this way.

My ham radio experience notwithstanding, I had an image of a scientist who was basically Einstein who (I imagined) could sit and contemplate the Universe and come up with a new theory of gravity. So understandably I harbored serious doubts about my ability to "cut it," to actually come up with new ideas.

But I did know I had a talent for learning. I took the science and mathematics courses offered by my high school as soon as I could. My high school teachers were very supportive at every turn. I doubled up on science whenever I had a chance, and took calculus in my senior year. But I knew that I wasn't close to knowing whether I could "be" a scientist. I knew I could study the books and pass the tests, but I wondered whether I could ever add to the books. Was I creative? I think proto-scientists torture themselves with this question as much as artists do. College would tell me, I hoped.

Losing my religion and heeding a call

In 1981, I accepted MIT's offer of admission and traveled halfway across the country to see if I could survive one of the toughest science programs in the world. I had been advised by my high school chemistry teacher, Sr. Mary Eva, that I was likely to lose my religion. Little did she know I was already well on my way; the God of my religion seemed more arbitrary than the Metric System. I had tentatively identified physics as a possible major, so I was paired with Dr. Bob Goeke, an engineer at the Center for Space Research. He had the classic nerd-look in those days; the thick glasses and the button-down white shirt. He seemed like one of those MIT genius guys. His second piece of advice to me (after I had politely declined to take a popular course in undergraduate mechanical design) was that, if I wanted to see what a physicist did, I should work for a physicist.

Up to then my work experience had consisted of minimum-wage farm and janitorial work, clerical odd jobs, and a brief unsuccessful stint selling dress shoes. The concept of working for a physicist blew me away, but a few weeks later, I took him up on it, and he set up a work-for-credit position with Drs. Tom Markert and Claude Canizares, in the X-ray astronomy group at MIT.

That job and those men have colored the path of my career ever since. I had no idea how significant this choice was; my instinct was to grab *any* job that (a) would pay better than minimum wage and (b) had serious air-conditioning. Even more importantly, Tom and Claude showed me what physicists could be. Tom was low-key, soft-spoken, and idealistic; Claude had the awesome white hair of an Einstein, but he dressed much better, and is one of the most classy individuals in astrophysics. I absorbed much by pure osmosis over the four years I worked in their group. I remember Tom describing my first project, sketching a rough diagram on a pad of wide-lined paper: the super giant explosion of a star, a supernova. Now that is an explosion!

X-ray astronomers study astronomical objects that emit X-rays. Such objects emit X-rays for usually one of two reasons: either they are very hot, as in the case of the supersonically shocked shells of interstellar gas left behind after a supernova, or they are composed of charged particles moving at speeds close to the speed of light—relativistic particles. Such particles exist near supermassive black holes, which can accelerate atomic particles to high speeds. X-ray astronomers must use telescopes that soar above the Earth's atmosphere, because the atmosphere absorbs the X-rays from the cosmos, luckily for us. X-ray astronomers thus must rely on NASA to fly X-ray telescopes to support X-ray astronomy.

Tom and Claude both studied the hot gaseous X-ray emitters, for example, the supernovae and the intergalactic gas between the galaxies in clusters of galaxies. My undergraduate career began with helping Tom analyze his supernova data from an early orbiting X-ray telescope called the Einstein Observatory (November 1978–April 1981), and ended with an undergraduate thesis on the Pegasus I cluster, based on data also taken with the Einstein Observatory. I decided, based on my undergraduate experience with my thesis, to be an astronomer, in the final weeks of writing it in January 1985.

I spent all four years as an undergraduate not knowing if I had whatever it takes to be a scientist. I was surrounded by people who seemed so certain of themselves. I had reasonable grades, but so did many other people. Tom and especially Claude seemed like people who operated on some higher plane, plucking ideas and associations out of the air as if by magic. I could execute the technical side of data analysis, but I knew that wasn't "it," the gift of doing original research.

But in January of 1985, I received a glimmer of what the experience of being a real scientist might be like. Up to that stage of the thesis I had been working on the data at the computer—cleaning it, filtering it, re-extracting it from the original data files, fitting the emission profiles. But when I began to translate my measurements and mathematical fits into what I could say about the cluster intergalactic gas itself, I got excited. The energy fluxes could be translated into real physical properties that I could write down: luminosities, electron densities, temperatures. The properties that I had measured for the contents of a gas discharge tube during my junior laboratory class could be measured from an ionized plasma a quadrillion miles away. I could estimate how much iron was in that gas, and how much cool gas lay in our own Galaxy between us and the cluster. All of this came to me in a rush as I reviewed the X-ray literature and realized that I had the "unknowns," the observations, required to complete the equations and reveal the physical properties of the cluster gas. The physics of the gas was revealed in its radiation, so measuring the properties of the radiation told me—told us—about the gas.

This revelation is nothing new for astronomers—they do this all the time, every day. But it rocked my world. I was learning about a cluster of galaxies no-one had ever observed in this way before. As I was writing down my results, I was very self-conscious that these were answers to questions no-one had ever answered before, not about this cluster, not in this way. Suddenly the work became not-work, and for my last semester at MIT I lost track of time whenever I worked on my thesis project.

I was hooked on astronomy and I never looked back after that. I even regretted applying to pure physics graduate programs. I had applied for graduate school in December and January to the most general physics programs I could, fearing unemployment as an astronomy graduate. I was now so certain that I wanted to go for an astronomy degree that I realized that applying to those physics programs was a waste of time, not to mention that I didn't even get in to most of them. They probably suspected I was an astronomer at heart. I was still not certain I was "scientist stuff," but at that point I had found a joy in the doing that was worth pursuing even if only for the next few years of my life.

Summoned to the mountains

X-ray astronomy at that time had rather bottomed out. There were no new NASA X-ray missions planned for at least the next five years. The

Hubble Space Telescope (HST) was in the near future; graduate students entering in 1985 could perhaps plan to do their PhD theses with HST data. I played a conservative bet by choosing the University of Colorado at Boulder. I was recruited by the theory group there, consisting of Drs. Mike Shull, Dick McCray, and Mitch Begelman. I figured theorists would always have work to do, regardless of the pace of NASA missions. I knew I could do data analysis, but could I contemplate the Universe and come up with a new theory? Probably not, but I could watch people who got paid to do that kind of thing every day, and maybe learn the secret.

Mike Shull, a *Wunderkind* with a quick wit and bouncy optimism, included me in his research group and immediately began tutoring me in a crash course on interstellar and intergalactic gas physics. These topics reappeared in his course on internal processes in gases, but he wanted to jump-start me into a research project involving, of all things, intergalactic explosions. Was I already in a research trend, studying ever more fantastic explosions? I had started by studying the explosions that could rearrange the planet; now we had moved on to far far more energetic events.

These explosions aren't really single events, but shocks driven by many supernovae going off at nearly the same time in the same galaxy. I dutifully studied the shock and fluid equations, but I floundered around, wondering where I was supposed to go with it all, and just what new results I could ever contribute. I watched, with irritated envy, as my new boyfriend, Mark Voit, also a Shull protégé, scrawled new equations on his blackboard with ease, while I scratched two lines on my paper, lines which usually devolved into caricatures of ARGH.

My doubts returned.

But several months later I solved a cosmological ionization-front problem with an analytic form, which Mike Shull believed was a paper-worthy result. We wrote it up and submitted it to the *Astrophysical Journal Letters*. The controversy of that paper, which asked whether quasars could ionize the intergalactic medium (it basically said yes, if there are enough quasars, which seemed fairly non-controversial to me), challenged my toleration for stress. I got the impression that, to be a theorist, you had to have a solid ego, robust self-esteem, and strong stomach because, if you didn't, you were going to get hurt. It was very easy to take the criticism personally.

By 1988, Mark and I had married. By doing so we had made pursuing careers in astronomy even more difficult. How to find two astronomy jobs in the same place, the infamous "two-body" problem in science. I decided that two theorists were less likely to find jobs together than would a theorist and an observer so I happily embarked on a separate thesis project with another advisor, Dr. John Stocke.

John, a recent Colorado hire from Arizona, as new-age relaxed as Mike was intense, was just completing one of the major projects of his early career: the identification of all of the extragalactic X-ray sources observed by the Einstein Observatory. X-ray astronomy was coming back into my life. John and his multi-national group of collaborators had re-analyzed all of the archived image data acquired by the "IPC" (Imaging Proportional Counter) detector on the Einstein Observatory to search for new X-ray sources. Usually, the target of each observation was located in the center of the image. But the Einstein Observatory IPC was a thousand times more sensitive than any previous X-ray experiment, and it was the first to make true X-ray pictures of the sky. New, previously unknown X-ray sources would appear in the image along with the original target. The position in the sky and the X-ray brightness of each source was cataloged by John's group; but little else could be known. The only way to find out just what these sources were was to look at their sky positions with other kinds of telescope, such as optical and radio ground-based telescopes. Nearby stars would show up very brightly in the optical part of the electromagnetic spectrum; exotic quasars often shine strongly in the radio.

These radio and optical data revealed the true nature of the unknown X-ray sources. Normal stars need to be relatively close to be detected in the X-rays; such sources were fairly obvious after even a brief inspection of the Palomar Sky Survey. But they are not the most common X-ray sources. The physical conditions required for gas to emit X-rays are among the most extreme conditions in the Universe: either very hot or very relativistic (very high speed). The X-ray catalog that Stocke and his colleagues constructed, called the Extended Medium Sensitivity Survey (EMSS), contains quasars, flare stars, binary systems where one of the companions is an exotic collapsed star such as a neutron star or black hole, supernova remnants, and clusters of galaxies.

The objects that caught my interest were the clusters of galaxies. Clusters of galaxies emit X-rays because the intracluster gas, the gas between the galaxies in the cluster, is compressed and heated by the enormous weight of the cluster. Clusters of galaxies have tremendous quantities of dark matter, binding the cluster together by its gravitational attraction. The total mass of the most massive clusters of galaxies is several million billion times that of the Sun, of which about 5–10% lies in the thousands of galaxies in the cluster and another 20% or so in the intracluster gas. The galaxies orbit around the center of mass in the cluster, flying around at velocities up to 1,000 km/s, and the intracluster gas is confined to the cluster by the same gravitational attraction. The gas particles move at essentially the same speed as the galaxies. Since the temperature of a gas is physically a measure of the random speeds in the gas, a gas whose particles are moving around at random speeds of 1,000 km/s has a temperature of 70 million kelvin, hot enough to emit X-rays.

I topped off my doctoral thesis on intergalactic and intercluster gas with a study of cooling flows, a special type of cluster of galaxies where the central gas seems to be too dense and too luminous to remain that way for the entire history of the cluster. A "flow" is inferred because such gas is expected to lose energy, and thus the core pressure decreases, allowing gas from the outer parts of the cluster to settle slowly in toward the center. Little evidence existed for a tremendous amount of matter piled up in the cluster core—no cold gas, no giant bursts of star formation. The physics seemed so simple, and many of the cooling-flow clusters, as they are called, showed such wonderfully exquisite and mysterious phenomena in their cores: extensive and colorful optical emission-line nebulae, which are usually associated with star formation, and peculiar radio sources, in the centers of elderly elliptical galaxies, which are usually giant collections of old stars, well past their star-formation days. Clearly, *something* unusual was happening in these central gal-axies; the question was what.

What I didn't appreciate was that this seemingly narrow field of inquiry was a focal point for some of the greatest personal passions I have yet encountered in astrophysics; in my limited experience, it is second only to the pursuit of the Hubble constant (a measure of how fast the Universe is expanding). In the field of cooling flows in the 1980s, you were either a believer or an unbeliever, and people were remarkably emotional about their beliefs. In most fields either you are very positive about the work, or you are relatively neutral. In astronomy, there are individuals who actively hate the topic of cooling flows. I was told that a review committee once reviewed my job application with the assessment "Too bad she wrote a thesis on cooling flows." The choice of research topic turns out to be very crucial in astronomy—that choice you make determines what places will consider you for employment— yet few students (myself included) recognize that at the outset of their careers.

After I completed my thesis in 1990, Mark and I were fortunate to find temporary postdoctoral positions in Pasadena, California. He went to work with Sterl Phinney in the theoretical astrophysics group at Caltech, and I accepted a fellowship at the Observatories of the Carnegie Institution of Washington, which at the time not only owned its own telescopes in Chile but also shared a fraction of the Palomar Observatories with Caltech and Cornell. I now had significant access to telescopes to follow up on my ideas about cooling flows, but I was also allowed to explore new avenues of research. Most importantly, at Carnegie I learned how observers "do it." For one of my thesis chapters, I was handed the core of an observational project, which I then developed and executed, and supported with calculations of new theoretical models. My unique contribution had mostly been in the area of theory. At Carnegie, though, I had to not only learn how to operate and calibrate the instruments, but also craft my experiments with judicious choice of sample, sample size, and experimental technique. Carnegie astronomers are free to choose their own research, and the permanent staff there have been able to embark on some of the greatest and most ambitious observing programs of the latter twentieth century.

Coming from a mostly theory-based background, I did no such thing. My experience there was like an apprenticeship, where I was able to experiment at the telescopes, with the instruments, but without worrying about generating immediate results. I did carve out some pointed experiments combining my theoretical work with some careful observations of single "cooling-flow" clusters. But I was impressed with the grand scope of the Carnegie science, and the seeds of what was to be my current research focus were planted. Cosmology was cool!

To a cluster very very far away

As a side project at Carnegie, I took electronic snapshots of the remaining unidentified southern X-ray sources from the EMSS. Some celestial objects, such as the Large and Small Magellanic Clouds and Alpha Centauri, are only visible from the southern hemisphere. So, while I was in Chile, I snapped images through the duPont 100-inch telescope of southern EMSS source positions with no obvious optical counterparts.

One night, as I took these images, I saw the most amazing sight on my computer screen. As the image I had just taken displayed itself, rather than the nondescript collections of stars and galaxies that I was used to seeing in these blank fields, I saw a dense bees' hive of little faint smudges: a distant cluster of galaxies. I could barely see these smudges through our reddest filter, and a subsequent image through a bluer filter showed no galaxies whatsoever. The fact that the cluster galaxies showed up through the reddest filter but not through a bluer filter meant that the galaxies were very far away—nearly halfway across the Universe. That was my first view of the cluster MS1054-0321.

Like meeting someone who would subsequently become your best friend years later, this cluster consistently appeared in my life after that night. A year later, I was observing on Mount Palomar, desperately trying to discover the exact distance to this cluster by measuring the spectrum of at least one of its galaxies. An image of a galaxy is relatively easy to get, because all of the photons more or less pile on top of each other, making a nice significant spot on the detector. But a spectrum is much more difficult to make, because the photons, the light, are dispersed in a line across the detector, rather than in a single spot. The dispersion of the light depends on the energies of the photons, so we can tell what energy the light has. A spectrum tells us much more about an astronomical object than an image does; in an image we lose information by stacking photons without regard to their energies. But a spectrum, in the bands of light sorted by energy, can reveal how fast the object is moving towards us or away from us, the elemental components of the gas and their ionization states, the temperatures of the gas and much more. Here I wanted to measure how much the Universe had "stretched" (expanded) since the time the light had been emitted. This expansion is also known as "redshift;" the redshift of the light emitted by the cluster galaxies would tell me how far away that cluster was.

I obtained an extremely noisy spectrum and was only able to estimate a redshift of about 0.8, which means that the spectrum of the galaxy was stretched by 80% (0.8) to longer wavelengths. Later, Isabella Gioia, from the University of Hawaii, took advantage of the remarkable skies above the Canada-France-Hawaii telescope and found a redshift of 0.83 for a few of the galaxies. With such a high redshift, the cluster was confirmed to be nearly halfway across the known Universe.

This distance was very significant, because in the standard model of cosmology and the formation of large-scale structure in the Universe, such clusters weren't supposed to be there. If this cluster was not only distant, but also hot and massive, its mere existence was going to be tough to explain in the context of these models.

X-ray astronomy had moved into the spotlight again in 1990 with the launch of a joint German/UK/US satellite observatory called ROSAT. ROSAT's window on the electromagnetic spectrum was too low an energy to enable us to measure cluster gas temperatures, but its excellent imaging detectors meant we could measure the shapes of the cluster, by taking a picture of the X-ray emission coming from the cluster gas. I began to apply for time to study some of the EMSS clusters in my thesis; but I was excited by a grand-scale project that Simon Morris, another postdoctoral researcher at Carnegie, was participating in to measure the velocities of galaxies in EMSS clusters with redshifts of 0.3 to 0.5. By measuring the velocities of these cluster galaxies, Morris and his collaborators could tell how much matter was really there in each cluster. Such measurements had never been done well before for clusters so distant.

I did not feel I could intrude on their project, but I could start a project of my own to observe the highest-redshift clusters of galaxies in the EMSS, the clusters with redshifts greater than 0.5, including my friend MS1054-0321. To test the waters, I applied to ROSAT to observe the most luminous cluster in the EMSS, MS0451-03 at redshift 0.55. When that proposal was awarded observing time, I conceived of a project that would study all of the high-redshift EMSS clusters. These were some of the only clusters known in the distant Universe. Because the distances to the clusters of galaxies is so immense, light takes a long time to get from these clusters to us on Earth. This time delay means we're seeing the clusters, not as they are now, but as they were billions of years ago, when the Universe was only about half as old as it is now. Clusters of galaxies are also very rare objects. Studying how samples of rare objects change with time was sure to be interesting and possibly relevant to cosmology.

With the support of my colleagues John Stocke and Isabella Gioia, we began applying for more observing time. We speculated that perhaps the lumpiness of these clusters would tell us something about how structure forms, and the density of the Universe.

The density of the Universe is a quantity that astronomers are rather keen to measure. If the average density of the Universe is the critical density or less, the Universe will continue to expand forever. If the average density is greater than the critical density, it will eventually stop expanding and collapse. The critical density is actually very low, the equivalent of a few hydrogen atoms in a space the size of a typical closet. The value Ω_m is defined to be the ratio of the average density of matter divided by the critical density. A critical density universe (or, equivalently, an $\Omega_m = 1$ universe) is a very appealing theoretical model because it is so simple, and has a plausible explanation in the form of the inflationary Universe, a key variant on the Big Bang theory of the Universe's origin.

Theoretical work by Doug Richstone and others suggested that, in a high-density universe, clusters would be as lumpy in the past as they are today. Optical observers, using galaxy positions and velocities as evidence, were jumping up and down at conferences saying that the Universe must be critical density, because even local clusters of galaxies were so lumpy. I thought that a better test must be to compare local clusters with distant clusters, since "lumpiness" is difficult to define, but comparing the lumpiness of clusters nearby with that of distant clusters might be easier to quantify. If distant clusters were *more* lumpy than they are today, that would support the idea of a less dense universe.

But I happened upon an even more compelling test of the density of the Universe in a comment by Monique Arnaud in one of her 1992 papers in the European journal *Astronomy and Astrophysics.* In this paper she and her colleagues describe one of the hottest (and therefore the most massive) clusters of galaxies ever discovered. It was so hot and massive, she explained, that its very existence in the volume of space where it was discovered was a challenge to the dense Universe ($\Omega_m = 1$) theory. She was not the first to say that the existence of a very massive cluster of galaxies was very difficult to explain in the context of an $\Omega_m = 1$ universe—Dr. Jim Peebles of Princeton University in his 1989 paper with Drs. Ruth Daly and Roman Juszkiewicz was one of the first—but she was the first to apply this concept to an X-ray cluster in an X-ray survey, and it was Monique's paper that planted the idea in my head of making this test.

In 1993, we were lucky again—Mark won a Hubble fellowship to go to Johns Hopkins University (JHU) and I was offered an Institute fellowship with equivalent prestige at the Space Telescope Science Institute, just across the street from the physics and astronomy department at JHU. We packed up our Siberian husky and moved across the country to Baltimore.

Yet another X-ray satellite was launched in 1993, jointly by the Japanese and the US. This satellite was named ASCA (Advanced Satellite for Cosmology and Astrophysics)—really a pun for an ancient Japanese word for "flying bird" or Asuka. ASCA was perfectly suited to take spectra of clusters of galaxies, and therefore was perfectly suited to take a cluster's temperature. I immediately applied to observe MS0451-03 in the first round and was rewarded with a first-priority, early mission observation!

Is the Universe dense? Clusters say no

In the summer of 1994, I spent three weeks at the Aspen Center for Physics discussing the exciting new cluster observations from the ground and from space. I had just completed my analysis of the MS0451-03 data. The cluster had proved to be rather hot; Keck observations of its galaxies by Isabella Gioia showed that the velocity dispersion (the distribution of galaxy redshifts around the average cluster redshift) was high too. I was also pregnant with my first child. I was feeling a little uncertain about my future in astronomy, but for the first time I was feeling like I was beginning to make a major contribution. Richard Mushotzky, a high-energy, bearded MIT graduate, a pundit in spectral X-ray astronomy, encouraged me to go for the more faint, distant clusters in the EMSS with ASCA. So I did so that summer and, later that fall, I wrote a proposal to NASA for a five-year project that would support my salary while I pursued these high-redshift clusters. That winter, Mark and I began applying for more permanent "tenure-track" positions. Tenure-track positions can last longer than typical 2–3 year postdoctoral research positions, although they have their own flavor of instability, in the form of the tenure review. But that is a fact of life at most academic sites. We were encouraged to apply to Space Telescope Science Institute (STScI), which was planning to hire at least three astronomers that year, an unusually high number for a single institution.

Both Mark and I made the "short list" at STScI, and, 9 months pregnant, I gave my job talk. Mike Shull happened to be sitting in the audience, and he was very pleased and optimistic. Mark gave his job talk a week later, and then, right on time, our daughter Michaela Voit was born. Within two months, we were relatively secure again. I had been awarded a long-term grant from NASA, and Mark was offered a tenure-track job at STScI. That summer, STScI also offered me a tenure-track job, which I happily accepted.

In 1995–6, I applied for ASCA and ROSAT time to study my EMSS cluster sample. Proposing was time-consuming work, because the clusters were so faint that each one nearly required its own proposal to get sufficient time to get the data I needed. I also worked on some projects in other areas, traveling to the telescopes at Kitt Peak occasionally and applying for Hubble Space Telescope time to study galaxy appearances in the distant clusters. In January 1997, I went back to Aspen, this time for a winter conference. I had my ASCA data for the most distant cluster in the EMSS, my MS1054-0321 cluster (Figure 7.1, color section). But I was having a little trouble fitting it into a theoretical context. I had measured a very high temperature for the cluster, which meant that it was very massive. But its very high temperature also meant that it was extremely rare—too rare for the cosmological simulations that had been published in the literature. When I looked to see how many clusters were predicted at redshift of 0.8 for such a hot temperature, the simulations simply did not have enough volume to contain even one cluster. That was not very satisfying!

At the Aspen meeting, Carlos Frenk, a noted expert in large-scale structure, had told me that such a cluster would really be the death knell for the dense Universe ($\Omega_m = 1$). Such a comment is like music to an observer's ears: a major theory that could be falsified by my data!

I worked on that paper through the spring of 1997 and finally submitted it in July 1997. I was by then pregnant with our son, but I hoped I could get the paper through the refereeing process before his birth. Unfortunately that was not to be. The paper was not refereed until November 3, over a month after Sebastian's birth. While we revised the paper, Mark came up with some theoretical insights, which we wrote up and submitted as a separate paper that was reviewed and accepted very quickly. We were also able to construct a simple analytic argument that circumvented the need for voluminous numerical simulations. Despite the delay, I was very happy with the way the paper turned out.

Our main result from the paper was that we found that MS1054-0321 was so hot that the chance of finding such a cluster at redshift 0.8 in an $\Omega_m = 1$ universe inside the volume of the Universe sampled by the EMSS was about 1 in 100,000. In fact, only one such cluster was expected in the entire Universe, so the chances of finding even one in a small sample volume is very small indeed. The idea behind this is rather simple. The formation of clusters is sensitive to the average density of the Universe. It turns out that if the average density of the Universe is the critical density ($\Omega_m = 1$) then clusters form continually, that is, once they collapse out of the general expansion, they continue to grow and grow forever. So, if $\Omega_m = 1$, the clusters today are, on average, more massive than clusters long ago.

But if the Universe is less dense than the critical density, clusters don't continue to grow so vigorously forever. At some time in the past (depending on the density), clusters significantly slow their formation and eventually stop growing at all. For the lowest-density models of the Universe, there is very little difference between how many massive clusters there are today and how many there were in the past.

So, if you know how many clusters exist today (nearby) with a certain mass, you can predict, very simply, how many clusters should have existed with that mass a long time ago, say, when the redshift is 0.8 and $\Omega_m = 1$. If $\Omega_m = 1$ and the mass is very massive, say one million billion times as massive as the Sun, like MS1054-0321, then the difference between the number of such clusters now and the number of such clusters back then is simply enormous. But if $\Omega_m < 1$, that is, if the Universe is much less dense than the critical density, there would be almost no difference.

What we saw was not just one such cluster at redshift 0.8, but two additional clusters in the EMSS that were nearly as hot and nearly as distant (redshift 0.5). So we were convinced that our observations weren't just of a one-shot rare, bizarre event. We had at least three clusters in hand that just shouldn't have been there if the Universe were really as dense as $\Omega_m = 1$.

I then went on to observe yet another cluster at redshift 0.8 to complete the EMSS sample. Our statistical and analytic analyses of the entire procedure for recovering the density of the Universe from measuring the numbers of clusters in a survey volume and their temperatures was recently accepted by the *Astrophysical Journal Letters*. We now have the result that $\Omega_m = 0.3-0.4$, depending on whether the Universe is actually accelerating or not (the lower value corresponds to the accelerating Universe model.) We had found that the Universe is open, and will expand forever. Our results can't tell whether the Universe is accelerating or not, but they are affected a little bit by whether it is or not. So we just have a piece of this cosmology puzzle, and it's a very interesting piece!

Despite this finding, the position where we're left, in the current popular consensus for the Universe geometry, is a little unsettling. If $\Omega_m = 1$ is not true, that means that we live in a special time when structure formation is slowing down from its vigorous activity of the past. If the Universe is not merely expanding, but accelerating like the recent cosmological supernovae results and the preliminary cosmic microwave background results suggest, than we're living in a special time when that acceleration is just now beginning to be important, and before the time when that acceleration will sweep most of the known Universe out of our sight. In some sense, the $\Omega_m = 1$ Universe offered a perfect universe where we live in no special time, a universe where curious civilizations can come and go and study other galaxies until the stars burn out forever. We still have a lot of work to do to understand the Universe in which we live.

Sr. Mary Eva said that I would lose my religion. But some days, when I contemplate the mind-boggling implications of even some of our simplest observations, I suspect that my religion has returned. It's not religion as Sr. Mary Eva would define it. It's a religion in which human beings, in reaching out to understand the tremendous cosmos in its full glory, can still stand in awe of it and its Creator. It's a religion where both men and women can play and contribute, where it really doesn't matter, in the long run, who you are. Science is a naturally human enterprise. We are still stuck circling an average sun remotely located in a relatively average galaxy. But we can still ask questions, ask whether the questions are the right ones, and pursue the answers. And, occasionally, we get a glimpse of what may be the answer. These moments, as ephemeral and fleeting as they are, provide me with the incentive I need to go on and try again. The best scientists may experience those moments many times. But you only need to experience one of those moments once to know it is all worth the effort.

I've spent most of my life to this date wondering if I had what it takes to be a real scientist. But now I realize the main ingredients I had all along were just the curiosity that I had as a child and a fascination with nature's beauty and power. Sometimes I think we're all born scientists; I was just born lucky too.

Suggested reading

J. Bennett, M. Donahue, N. Schneider, and M. Voit, *The Cosmic Perspective*, Addision-Wesley Publishing Co., 1998

Malcolm S. Longair, *Our Evolving Universe*, Cambridge University Press, 1997

Dennis Overbye, *Lonely Hearts of the Cosmos*, Little Brown & Co., 1999

Vera Rubin, *Bright Galaxies, Dark Matters*, The American Institute of Physics, 1996

8

Dark matter and the discovery of galactic halos

JEREMIAH OSTRIKER, Princeton University

Jeremiah Ostriker received his doctorate in astrophysical sciences from the University of Chicago, under the tutelage of the legendary astrophysicist and Nobel laureate, S. Chandrasekhar. After receiving his doctorate, Jerry held a postdoctoral position at Cambridge University. He then went to Princeton University, where he became the Chair of the Department of Astrophysical Sciences and the Director of Princeton University Observatory. Since 1995 he has served as the Provost of Princeton University, while maintaining his position as a professor in the Department of Astrophysical Sciences. Jerry Ostriker's contributions to astrophysics have earned him the recognition of his colleagues in awards as diverse as the Helen B. Warner Prize and the Henry Norris Russell Prize of the American Astronomical Society, the Vainu Bappu Memorial Award of the Indian National Science Academy, and the Karl Schwarzschild Medal of the Astronomische Gesellschaft of Germany. Jerry's research spans much of the field of theoretical astrophysics, with his current interests focusing on cosmology. For this book, he chose to write about one of astronomy's longest-standing mysteries: the dark matter that pervades galaxies.

By now most of even the lay newspaper-reading public has heard of "dark matter." Where is it? How much of the stuff exists? What is it? And, incidentally, how sure are we of its presence, or could the whole scientific story for its existence collapse?

There is a wonderful quote attributed to Mark Twain after listening to a lecture by a famous astronomer: "I am always impressed at what a rich return of speculation can be obtained from such a small invest-ment in fact." Does the remark apply here? Actually, the discovery of dark matter in the Universe is quite the opposite. We were steadily bombarded with facts attesting to the existence of large amounts of dark matter, facts that we were not prepared to recognize.

What gravity reveals

It has been more than two thirds of a century since the first persuasive arguments were adduced for some form of pervasive dark matter by the extraordinarily creative (and irascible) Swiss astronomer Fritz Zwicky, but it is only during the last half decade that most cosmologists have come around to agreeing on the essential facts: stars make up only about 1% of the mass density of the Universe; gas left over from the Big Bang that has not (yet) condensed into stars (in galaxies) is perhaps ten times more abundant and makes up 10% to 15% of the matter density; the remaining 85%–90% is some strange "dark matter" of unknown composition. The stuff is probably "non-baryonic," that is, not made of hydrogen, helium, magnesium, iron or any of the other familiar chemical elements, and it probably does not interact significantly with ordinary matter in any fashion except one: Newton's and Einstein's laws of gravity cause all matter to pull on all other matter with a force proportional to the amount of matter but inversely proportional to the square of the distance. Thus, "dark matter" obeys the same laws as those that make the Moon orbit the Earth, the planets circle the Sun or the Sun circle around our galactic center once every several hundred million years (roughly the time since the extinction of the dinosaurs on Earth).

It is, in fact, via gravity that dark matter was detected—again and again—so the term "missing matter," which is fortunately falling out of use, was really a preposterous misnomer. It was as if, when you went to the bathroom scale and weighed in at 150 pounds, you always responded, "But really, I only weigh 15 lb; something is wrong here!" And you did this day after day.

The story begins with Zwicky, who (in 1933) discovered numerous clusters of galaxies in the extragalactic Universe. They looked like clusters of stars, so he used the same techniques to analyze them as had been used successfully for star clusters. One measures the velocities of the different stars, finding that the cluster is neither expanding nor contracting, but rather the velocities are random, with the stars orbiting around and through cluster centers like bees around a hive. Then, since the only force that can keep the stars from flying away from one another is gravity, one can compute the average gravitational force pulling the stars back towards the cluster center. Finally, since this force is produced by the stars themselves, one can deduce the mass of the cluster and, dividing by the number of stars, the typical mass of a cluster star. This is the same method, in principle, used to weigh the Sun using planetary velocities, and, reassuringly, for a typical cluster, like Omega Centaurus, it gives a mass for the average cluster star quite close to the mass of our Sun.

But, to Zwicky's astonishment, when he performed this same exercise for the Coma cluster of galaxies, he obtained a mass one hundred times more than had been estimated from measuring the light emitted by the cluster and *assuming* that the ratio of mass-to-light for everything in the cluster was the same as that of our Sun. We now know that the Coma cluster is representative of the Universe; the stars constitute about 1% of its total mass, the hot gas between the galaxies perhaps 10%, and dark matter is the rest. Zwicky was not particularly happy or certain about his revolutionary conclusion, but no-one at that time or since then has been able to find a flaw in his reasoning. The result stood for decades, reproachfully assailing astronomers who could neither refute it nor accept it.[1]

During the 1960s, when I was in graduate school, the normal method of evaluating the mass of galaxies was to (again) use the motions deduced in flat spiral galaxies from the gas and stars circling in the flat plane around a central mass. In our Solar System, where the mass is concentrated in the Sun, the measured velocities fall off, following Kepler's laws, that is, by the square root of the distance. So Jupiter, five times farther from the Sun than the Earth, has an orbital period of almost 12 years and an orbital velocity smaller by a factor of just over 2. But, when observers made the same kind of measurements in our and other galaxies, they found a puzzling and contrary result. The orbital velocity would rise as one moved outward from the galactic center, and

---

[1] Zwicky was equally prescient in other areas and foretold the existence of both neutron stars and gravitational lenses, long before they were discovered.

then it would remain *constant*, that is, the rotation curve would be "flat" and not show the expected Keplerian decline. Case after case was studied and none showed what we "knew" to be the case.

I read these papers as a graduate student, noticed that the observations (all of the relatively bright inner parts of galaxies) did not at all fit the models which the observers were using to analyze the data. Clearly, something was amiss. I knew of Zwicky's result and wondered if it was relevant.

Then, for my thesis work at the University of Chicago with Subrahmanyan Chandrasekhar and my early postdoctoral work with Donald Lynden-Bell in Cambridge, I worked on the classic problem of rotating, self-gravitating bodies: what are the allowed shapes and at what degree of rotation do axisymmetric objects become unstable and transform themselves to binaries or bars?

What I found was that the classical work on uniform, incompressible stars, started with McClaurin (1742) and culminated with Chandrasekhar, was easily generalized to compressible objects and might even be true for stellar dynamical systems like galaxies. If the ratio t, of rotational kinetic energy T to gravitational energy $|W|$, $t = T/|W|$, was greater than about 0.14, it seemed that both stars and galaxies might be unstable to the formation of bars. This work, often performed via numerical simulations on early "supercomputers," continued through the early 1970s. Then, in a 1973 paper with Jim Peebles, "A Numerical Study of the Stability of Flattened Galaxies: or, Can Cold Galaxies Survive?", a test was made that indicated galaxies in fact could not survive if they were as flat as they appeared to be. However, an "invisible" quasispherical halo would save the day, and we proposed that such a halo of dark matter might exist, within the observed galaxies, that acted to stabilize them.

Emboldened by this result, we looked at the observational data in a 1974 paper, "The Size and Mass of Galaxies, and the Mass of the Universe" (J.P. Ostriker, P.J.E. Peebles and A. Yahil), and concluded that both the size and the mass of galaxies had been greatly underestimated (by approximately of a factor of 10) and that, after correcting for this, the mass density of the Universe was much larger than had previously been estimated.

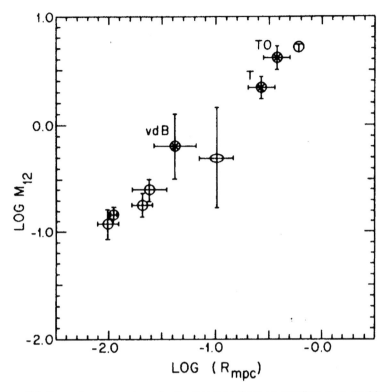

Figure 8.1

This figure showed, perhaps for the first time (from J.P. Ostriker, P.J.E. Peebles & A. Yahil, [1974] *Astrophysical Journal*, Vol. 193, Part 2), how the mass of a typical galaxy like our own seems to keep growing as one measures the total mass within spheres of ever-increasing radii. On the vertical scale, 0.0 represents 1,000 billion solar masses, −1.0 represents 100 billion solar masses (the conventional "mass of the galaxy"). On the horizontal scale, −2.0 represents approximately the solar orbit around the galactic center, −1.0 is a factor of 10 farther out and −0.0 is a factor of 10 still farther out.

In this second paper, we took an empirical, not a theoretical, approach. That is, we looked at all the different means that had been used to weigh our own galaxy (surely the best studied case) and assembled the evidence on one figure (Figure 8.1). In the inner parts, we relied on rotation; at intermediate distances, we used the fact that satellite systems were tidally shorn by the gravitational force from our galaxy.

At the largest scale, we noted the fact—previously analyzed by Lo Woltjer and Fritz Kahn—that our neighbor galaxy, Andromeda, was approaching us. This is not expected in an expanding universe and could most easily be understood if our local group (including both the

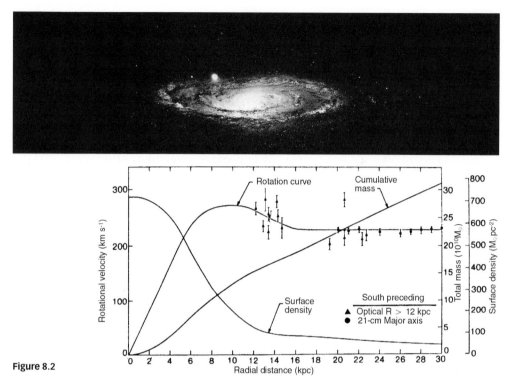

Figure 8.2

(*Upper*) Our companion galaxy, the Andromeda Galaxy (type Sb). (*Lower*) On the same scale, the rotation velocity of gas orbiting its center and, derived from that, the total mass within spheres of increasing radii (27, 28). Note that the mass in the outer parts continues to increase in regions from which very little light is emitted, implying that most of the mass is not in ordinary (solar) type stars, but some other dark form. (J.P. Ostriker, [1977] *Proc. Natl. Acad. Sci. USA*, Vol. 74, No. 5, p. 1767.)

Galaxy and Andromeda) was much more massive than had been thought. Putting all the arguments together, we found that, as one moved from the inner 10,000 parsecs (our distance from the galactic center) out by a factor of 10 or 20, the total mass increased by a factor of 10, even though most of the light emitted by our Galaxy was from stars inside of the solar orbit (see Figure 8.2). The total mass we estimated for the Galaxy—about 2,000 billion solar masses—is still close to the current best estimate, and the corresponding density that we found for the Universe (about one fifth of the magical "critical value") is also close to current best estimates. This work, together with the studies it followed by Fritz Zwicky and along with the work that it was in turn followed by—Vera Rubin's work on galactic rotation curves—was instrumental in leading to the widespread acceptance of "massive halos" and the belief that "dark matter" was a major component in the Universe.

The reception which our work received in the 1970s, however, was frosty indeed! We were accused of every scientific sin in the book: overstating conclusions, ignoring contrary evidence, invoking hypo-theticals, etc. The fact that so many supporting lines of argument led to the same conclusions was no help: "Many bad arguments do not equal one good argument" was the rhetorical rejoinder. And it was certainly true that no single piece of evidence was truly irrefutable—from Zwicky's work on clusters in the 1930s, to ours on the Galaxy in the 1970s, to Vera Rubin's exquisitely flat extended rotation curves in the 1980s (which mirrored Horace Babcock's astonishingly flat rotation curve for Andromeda—his thesis published in 1939). But, as usually happens in such cases, the evidence accumulated, the contrary evi-dence—where it existed—evaporated, and the consensus gradually shifted.

Greater puzzles

Well, where does this leave us? As ever, with greater puzzles at the end of our study than we had at the beginning. First of all, what *is* the dark matter and is there only one kind of dark matter? For simplicity, we have been assuming that there was only one kind, and we needed to find "it." Now it appears that there are at least two types. Recent experiments made in deep mines, analyzing neutrinos emitted by the Sun, are best interpreted by a model in which the very weakly interac-ting neutrino has a small mass. But, if this estimate is true, and if the abundance of neutrinos is what straightforward theory indicates is likely, then the neutrinos do constitute "dark matter" but, at most, a few percent of the total amount, and some thing or things constitute the rest of it. What about dark stars—low-mass stars emitting so little light that we have not detected them easily—or planets, or cold dust, or hot gas, or anything else made of ordinary ("baryonic") matter? Signifi-cant amounts of all of these components have surely been found, and we may be surprised in the future by new discoveries; but, as of now, all these baryonic components fit neatly into the budget number given earlier—ten times the luminous stars, but only one tenth of the total mass density—established for baryons.

This last number is derived from standard calculations for the nuclear "cooking" of material in the Big Bang. The recipe says that, if you can measure the relative amounts of various light elements and isotopes (normal hydrogen, deuterium, helium, lithium, beryllium and boron), you can deduce the total baryon complement. The measurements and theory are within reasonable agreement and indicate the number given: all baryonic components should add up to perhaps 10% to 15% of the dynamically detected matter and only 3% to 4% of the critical value.

While this picture is internally consistent, it leaves us with an enormous dilemma. Most of the dynamically detected matter in the Universe is of unknown nature and origin. Whether this is a profoundly unsettling state for our knowledge or a peculiarly satisfying one will differ from person to person. This particular person, for one, finds it quite delightful to leave to successors appropriately grand, totally unsolved problems in cosmology . . .

9

Hunting the elusive invisible galaxy

GREGORY BOTHUN, University of Oregon

Greg Bothun is a northwesterner, educated in Washington State, briefly a professor at the University of Michigan, and now a long-time professor of astronomy at the University of Oregon. Greg, nicknamed "Dr. Dark Matter" by his friends, is interested (when not raising his two sons, hiking, playing softball, or golfing) in galaxy evolution and studies of large-scale structure in the Universe. In what follows Greg takes us on a very special journey that he traveled, to find the dim, lurking giants of galactica, the so-called low surface brightness galaxies.

Introduction

One of the assumptions in cosmology is that, no matter where you go in the Universe, the stuff you see when you get there is the same stuff that you already knew about. This is known as the Cosmological Principle. This principle asserts that the Universe, at any given epoch in its history, is homogeneous. Thus all observers should measure the same characteristics and same physical laws, independent of their exact location in the Universe. If this were not the case, then the Universe would be an arbitrary place and there would be no guarantee that, for instance, the law of gravity that holds in New Jersey would be the same as that which holds in California.

Much of observational astronomy is about detecting and classifying the stuff that is out there. For the first 50 years of this century, that task was devoted to stars. The principal result of that is the construction of the Hertzsprung–Russell diagram that shows the kinds of evolutionary states that stars can occupy. Application of the Cosmological Principle to this situation tells us that, if we lived in the Andromeda Galaxy, then we should be able to construct the same Hertzsprung–Russell diagram. That is, the kinds of stars and their evolutionary patterns are the same in Andromeda as they are in the Milky Way. With modern instrumentation, we can now verify this assertion by direct observation of the stellar content of Andromeda.

The latter half of the twentieth century has seen intensive effort to detect and catalog different kinds of galaxies. But detecting whole galaxies, themselves full of stars, is different than detecting individual stars in our own Galaxy. Stars are point sources while galaxies are extended sources. The difference between a point source and an extended source is easy to understand if you visualize your detector as consisting of individual pixels (a pixel is a picture element—your television screen is composed of up to 1,000,000 pixels or dots). A point source means that the light from the object essentially falls in one pixel while an extended object might encompass several hundred or thousand pixels. The detection of a point source depends only on the ratio of incoming light to detector noise in that pixel. For extended sources, however, there is an additional source of noise that the signal must compete against and that is the background night sky. Because the night sky is not infinitely black (between the stars), it has finite brightness and therefore noise (some parts are brighter than others). Thus, a galaxy must be detected against this noisy background of finite brightness and this creates a detection problem whose severity, until recently, was not properly appreciated.

This detection problem can be put in perspective as follows:

Suppose that you were interested making a catalog of iceberg shapes and sizes. To do this, you stand on some stable shore in the Arctic and observe icebergs. Iceberg detection consists of observing an iceberg sticking up above the surface of the water. Now suppose that the average height of the waves was 3 meters. This would represent a source of "noise" in the background. Icebergs whose intrinsic height above the water exceeded 3 meters would be easily detected and cataloged, while those whose intrinsic height was say only 1 meter above the water surface might be difficult to detect against the waves (a constant source of noise). So you make your catalog of icebergs under these observing conditions and you deliver your results at the annual Symposium on Iceberg Shape and Sizes. In fact, you have developed an entire evolutionary theory for iceberg formation and

Figure 9.1 High-contrast galaxies: M100 and M51. These are the galaxies most easily discovered and they populate our current catalogs.

melting based on the catalog you have constructed. After you deliver your scintillating lecture a voice from the audience asks "How do you know that most of the mass in the iceberg population isn't in small icebergs that can't be detected amidst the waves?"

Upon reflection, you realize that you don't know the answer to this question because you would require better observing conditions to test this hypothesis. So you go back to the Arctic and are lucky enough to observe one day when the water is perfectly calm (the background noise has been removed). If a population of small icebergs exists, you should now see it and therefore provide a definitive answer to the question.

The point here is that the presence of background noise serves to bias the kinds of objects that can be detected. In particular, only objects of relatively high contrast with respect to the background will be detected. Examples of such high-contrast galaxies are shown in Figure 9.1.

But do such beautiful spiral galaxies represent the typical kind of galaxy that is present in the Universe or are these just the largest icebergs that are the most easily detected? Certainly galaxies of lower contrast are not easily detected. Thus, in principle, very diffuse galaxies might go undetected as they are essentially invisible, that is, indistin-

Figure 9.2   Examples of very diffuse galaxies, previously undiscovered.

guishable from the noise of the night sky. Their recovery from the night sky then requires very careful observations which serve to suppress the noise (as much as possible). In essence, this defines the hunt that I have been involved with since 1984. The target is the elusive invisible galaxy. Are there scores of diffuse galaxies that effectively use the noise of the night sky for camouflage? The answer can only be found by mounting an expedition to specifically search for these objects. This is the story of that expedition which has resulted in the discovery of galaxies so diffuse they are barely discernible, yet they are out there, in large numbers (Figure 9.2).

The initial quest

On October 1, 1981, I started my career as a professional astronomer at the Harvard-Smithsonian Center for Astrophysics, located in Cambridge Massachusetts just outside of Boston. I had just finished my PhD thesis at the University of Washington about the properties of spiral galaxies in clusters of galaxies. The data in that thesis represented about 100 nights of optical observing and 50 days/nights of radio observing. Hence, early on I realized that much of astronomy was spent staring at the sky. It was during those staring episodes that I first started thinking about issues of sky brightness and galaxy detection. In fact, this thinking was reinforced whenever I was doing infrared astronomy as that is

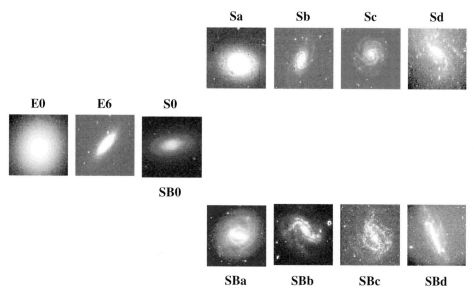

Figure 9.3   The Hubble sequence as defined by some representative high-contrast galaxies.

usually done during full moon time. With the optical sky lit up by the scattered moonlight I began to think of other observers who might live on a planet which had two moderate-size moons in orbit about it, at least one of which was in the night sky at all times. Under such conditions there would be no "dark" time and observers would be hard pressed to discover external galaxies, let alone even the thin band of stars we call the Milky Way. Clearly then, galaxy detection depends on the observing environment.

Also at this time I was immensely frustrated with the tools that astronomers used to understand the properties of galaxies. These tools largely consist of galaxy classification and morphology which has resulted in the Hubble sequence of galaxies, which is usually represented by a "tuning fork" diagram, such as the one in Figure 9.3.

One of the basic results I had found in my PhD thesis work was that quantitative measurements of spiral galaxy color (optical and infrared) and gas content did not correlate very well with morphological type, despite the fact that it was adopted lore that such correlations exist. As physical galaxy evolution is defined by the particular way in which a galaxy turns gas into stars, I thought that morphology was a poor substitute for physics and that the entire Hubble sequence might well paint a misleading picture of galaxy formation, evolution and structure. As I published these results I gained somewhat of a reputation as a renegade that was railing against the established way in which galaxies are characterized. Well, that was certainly true, but at the time I really didn't think that the best way to prove my point was to discover galaxies that were so diffuse that they couldn't be classified. After all, if you can't see (detect) a galaxy, you certainly can't classify it!

During my tenure (1981–3) at the Center for Astrophysics (CfA) I began to work closely with John Huchra and Margaret Geller on large-scale structure (see Huchra's chapter in this book). These were the early days of galaxy redshift surveys. Eventually these surveys would produce startling results about the cellular nature of the galaxy distribution and reveal to us a universe that was filled with large regions devoid of galaxies. To obtain a redshift for a galaxy meant obtaining an optical spectrum, usually with the 1.52-m telescope at Mt. Hopkins. Galaxies were selected for redshift measurement by consulting the Zwicky catalog of galaxies, produced by the irascible Fritz Zwicky using the Schmidt telescope at Mt. Palomar. This catalog is extremely biased toward containing only high-contrast galaxies.

Earlier, while I was acquiring the observational data for my thesis, I had noticed that in some galaxy clusters there were fairly diffuse-looking spirals that, in fact, were not contained in the Zwicky catalog, although many of them were contained in the Uppsala General Catalog of Galaxies painstakingly produced by Peter Nilsson at Uppsala University in Sweden (who deserves some kind of medal for his effort). In contrast to the Zwicky catalog, this catalog was based not on galaxy magnitude, but on galaxy diameter, so it did contain some examples of relatively diffuse galaxies. However, it's quite difficult to obtain an optical spectrum of a diffuse galaxy (because the galaxy signal effectively competes with fluctuations in the sky brightness) and therefore the redshift survey was biased against including such objects. Well, if the number of diffuse galaxies was the same (or larger) than the number of high-contrast galaxies, then we might have a severe violation of the Cosmological Principle.

Indeed, an important corollary to the Cosmological Principle would assert that all observers in the Universe should construct the same catalogs of galaxies. If this were not the case, then different observers might have biased views and information about (1) the nature of the general galaxy population in the Universe, (2) the three-dimensional distribution of galaxies, and (3) the amount of baryonic matter that is contained in a galactic potential. If we are therefore selecting only one kind of galaxy for redshift observation, we may end up with a biased view of the overall galaxy distribution. Put another way, astronomers have only cataloged the most conspicuous galaxies; those which exhibit the highest contrast with respect to the night sky background. It is these galaxies which define the Hubble sequence. Yet can we be sure that these galaxies are representative of the general population? I know that I wasn't so sure about this and my insecurity over the issue was the source of many robust discussions with doubting colleagues in those years. Of course, the only way to be sure was to prove my point that diffuse galaxies exist in very large numbers and have been a major component of the galaxy population that has been missed to date.

Alone in the wilderness?

Often times in science when you have an "opinion" (in an arrogant moment I might call it "insight") which is against the mainstream you are temporarily cast off into the wilderness. Ultimately, if you can't make the case, based on real data, for your opinion, you will simply remain in that wilderness. Being lost in the wilderness is strong motivation for finding others that might also be lost.

In the fall of 1983 I changed jobs (a frequent occurrence for a young astronomer) and took a position at the California Institute of Technology, in Pasadena, California. While I was happy to be back on the coast over which the Sun set, my first few days at Caltech were somewhat intimidating as this was another institution steeped in tradition that might take a dim view of renegades. During my first few days there, I met a fellow renegade and wilderness wanderer by the name of Chris Impey. He was a British astronomer (but I forgave him for that) and we became collaborators on what would turn out to be a long march out of the wilderness.

Impey had arrived at the wilderness by working on the properties of quasar (quasi-stellar object: QSO) absorption line systems. These enigmatic systems occur whenever cold clouds of gas are in the line of sight between us and a distant QSO. Often, there was no optical identification to these cold clouds of gas seen in absorption. I began to wonder if such absorption line systems could be due to a plethora of very diffuse galaxies, which, by my few observations, generally had a lot of gas. One night over a raging campfire in the wilderness (otherwise known as the John Bull Pub in Pasadena), Impey and I discussed this idea at length. He pointed me to a paper written in 1976 by Mike Disney, a Welsh astronomer. Disney's paper was largely mathematical in nature as he showed how, in theory, the finite brightness of the night sky acts as a visibility filter which, when convolved with the true population of galaxies, produces the population that appears in catalogs. Hence, all observers' catalogs will be biased toward detecting objects above some threshold contrast with respect to the sky background. Certainly, I thought, some of Disney's argument should apply to observers on this planet. After reading that paper, I was quite convinced that we did not have a representative sample of galaxies and I tried to drag Impey along with me in this conspiracy.

Discussions with prominent Pasadena astronomers of the time convinced me that they were in a comfort zone, dismissing Disney's original hypothesis as, at best, applying to a limited and inconsequential population of objects. I kept saying to myself, how can you be so sure? Much of the comfort seemed to be based on a paper published in 1970 by the Australian astronomer Ken Freeman. In this landmark paper, Freeman asserted, based on a carefully chosen sample of galaxies, that spiral galaxies exhibited a constant level of central disk surface brightness, namely $I_0 = 21.65 \pm 0.3$ magnitudes per square arcsecond in the blue part of the optical spectrum. This near constancy of I_0 has become known as Freeman's Law, and it became a natural target, as renegades are known for breaking laws. Under Freeman's Law, there was no room for a significant population of diffuse galaxies. As we will later show, the actual discovery of very diffuse galaxies, in relatively large numbers, will show that Freeman's Law was in error by approximately a factor of 1 million! That is, the now observed space density of very diffuse galaxies is approximately 1 million times higher than Freeman's Law would have predicted.

The essence of Disney's mathematical argument was that Freeman's Law was an artifact of selection and there does exist a population of large galaxies ("crouching giants") that are lurking just below the level of the night sky brightness. In more quantitative terms, the light distribution from a spiral galaxy is exponential in nature. The surface brightness at any radius r is given by

$$I(r) = I_0 e^{-r/a}$$

This characterization has two parameters: the central surface light intensity (or central surface brightness hereafter called I_0) and the scale length. The scale length (a) is the scale over which the light distribution falls to $1/e$ of its central value. One scale length encloses 26% and four scale lengths enclose 90% of the total luminosity of a disk galaxy. These two parameters determine the total mass and luminosity of the galaxy. The central surface brightness, I_0, is measured in units of magnitude per square arcsecond. On a moonless night in the blue part of the spectrum, the brightness of the night sky is 22.5–23.0 magnitude per square arcsecond. This means that, if you had a circular aperture of area one square arcsecond and pointed that at a piece of "blank" sky, the flux through that aperture would be equivalent to a 22.5–23.0 magnitude star. This background defines the finite brightness of the night sky as observed from the Earth. It is not infinitely dark (unfortunately).

Mike Disney's mathematical argument suggested that the typical galaxy which is detected would have a value of I_0 approximately one magnitude brighter than the sky background and this is exactly what Freeman found. More diffuse galaxies must exist, Disney reasoned, but they were just more difficult to detect. Given that the observed distribution of I_0 as found by Freeman was so tight, many astronomers thought it not possible for Disney's crouching giants to exist and hence they remained in the comfort zone. However, in 1987 Impey and I would blow this comfort zone right out of the water through the serendipitous discovery of the most massive (and luminous) spiral galaxy every detected, even though it's invisible. In fact, it was likely that that discovery also helped to rescue Mike Disney from the wilderness as well. But we are getting ahead of ourselves . . .

Commencing the hunt

Freeman's law showed that I_0 was represented by a Gaussian distribution without much variation around the mean value (this is called a narrow scatter). In a Gaussian distribution, the probability of finding an event that was more than 4 dispersion units from the mean is approximately 1 in 10,000. In terms of surface brightness, this translates into spiral galaxies which would have I_0 fainter than 23.0 magnitudes per square arcsecond. Well, my own 1981 thesis data showed the presence of a half dozen spiral galaxies below this limit in the Pegasus I cluster of galaxies. If Freeman's Law was correct, then there should be 60,000 other spiral galaxies in the Pegasus cluster that were of higher surface brightness. Of course there weren't, there were only about a dozen or so higher. However, this was a thesis result and the most important part of a PhD thesis is getting it done, not thinking deeply about the results. So, I missed this one entirely and would have to rediscover it about five years later. So much for insight.

Recall that in my days at CfA I was worried that the CfA redshift survey would return a biased view of the galaxy distribution. In 1984, I conducted my first observational experiment to test this conjecture. Using the Nilsson catalog, I produced a sample of approximately 1,400 diffuse galaxies which did not have measured redshifts at the time. Approximately 500 of these were in the declination range accessible to the Arecibo radio telescope. In what has to be the most hectic observing run I have ever had, we observed this sample 24 hours a day for eight days and detected about 65% of the galaxies in the 21 cm spectral line of neutral hydrogen. The most important outcome of this observing run, besides the side medical experiment on how to observe for eight days with no sleep, was the acquisition of a significant number of new redshifts of very diffuse galaxies. From these redshifts, approximate distances to the galaxies could be obtained. Knowing the distance allows for an estimation of the intrinsic size and mass of the galaxy. We found that the majority of these diffuse galaxies, in fact, had masses and sizes that were as large or larger than our own Milky Way.

Before this study, small catalogs of diffuse galaxies had been produced, most notably by Sydney van den Bergh in the 1960s working at the Dominion Astrophysical Observatory in Victoria BC, Canada. However, the objects in those catalogs all turned out to be very nearby and very small galaxies, called dwarf galaxies. While these are fascinating objects in their own right, they were not the same kind of objects that we were discovering in our surveys for diffuse galaxies. So the Arecibo redshift survey was the first means of establishing that there can be galaxies as big and massive as the Milky Way or Andromeda, but are nonetheless quite diffuse. In fact, the middle galaxy in Figure 9.2 is an example of a very large and massive spiral which is nonetheless quite diffuse and not the beautiful kind of spiral seen in Figure 9.1. So the initial hunt was successful; we had found diffuse galaxies of the size of the Milky Way and more than one of them! If these are a new class of galaxies, then they need a new name. Borrowing on a 1983 paper by William Romanishin and Steve and Karen Strom, which also explored the properties of a small sample of diffuse galaxies, we called this new class low surface brightness galaxies (or LSBs). However, as a sign of inertia in the world of astronomy textbooks, the term low surface brightness galaxy is only now just appearing, almost 20 years later.

Intensifying the hunt

In early 1984, the renowned astronomer Allan Sandage of the Carnegie Observatories published some of the first results in his Las Campanas Photographic Survey of the Virgo Cluster. Contained in those papers were some vivid examples of galaxies in the Virgo cluster, which were extremely diffuse. In contrast to our Arecibo sample, the diffuse galaxies in the Virgo sample of Sandage had to be dwarf galaxies, if indeed they were members of Virgo. Given their extreme diffuseness, I became interested in the question of how such apparently fragile galaxies could maintain themselves against tidal forces within the cluster. In clusters of galaxies, individual members often have grazing encounters with other galaxies which produce a tidal force in the galaxy in the same way that the changing Earth–Moon–Sun angle produces tides on the Earth. These grazing encounters between galaxies are often sufficient to remove stars and gas that are located in the outer, less-dense regions. But since LSB galaxies were everywhere of low density, it seemed probable that such encounters would completely destroy them, yet they seemed to flourish in the cluster environment. Furthermore, I wanted to know what kind of stellar populations must be present that produce light but result in a galaxy that you can barely see? Note that I still don't know the answer to this question!

Despite the high degree of quality control in the Sandage survey, Impey and I wondered about the possibility that this survey had missed galaxies of even lower surface brightness. To investigate this we enlisted the help of David Malin in Australia. David Malin was well known (and still is) to be a photographic wizard. He developed a method of photographic amplification that would bring out very-low-contrast features. David agreed to apply his technique to some selected one square degree areas of the Virgo cluster for myself and Impey, even though we offered him no guarantee that we would find anything that Sandage hadn't already seen. Nonetheless, David Malin is a very jovial and collegial astronomer, and he sent us very-high-contrast prints which showed evidence for some really diffuse objects. Impey and I excitedly circled these "little buggers" and showed them to our colleagues who nicknamed our collection of circles the "smudge" galaxies. Most of our colleagues scoffed at the notion that these were extremely diffuse galaxies and the most common claim was that our "buggers" were either, dust, water spots or some artifact of the photographic amplification. In other words, we were told that these faint and diffuse objects were not real.

Now we had an arduous chore before us. We had to convince Telescope Allocation Committees (TACs) that (a) we weren't crazy and (b) we really wanted to use CCD (charge-coupled device) imaging to verify that the smudge galaxies we had seen on Malin's film were real. Either out of sympathy or a desire to make us go away, TACs provided us a generous allotment of telescope time at the Las Campanas 100-inch telescope in late 1985 and early 1986. CCD follow-up imaging of the "smudge" galaxies proved to be highly successful as all of them were real. Not a single water spot, speck of dust, or artifact among them—just real

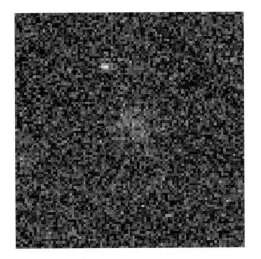

Figure 9.4

A digital smudge galaxy—one of the Malin "artifacts" that turned out to be real.

galaxies that you could see right through for the most part, a good example of which is shown in Figure 9.4.

Most of the smudge galaxies we had found turned out to be rather devoid of structure and were basically diffuse blobs. But they were real and had been missed by the Sandage survey. Furthermore, one of these objects had what looked like a very faint spiral structure which was connected to a point-like nuclear region. It was distinct from the other smudge galaxies in that it had what appeared to be a distinct nucleus that might be bright enough for an optical spectrum. On the Palomar Sky Survey this nuclear region is visible as a faint star but no associated nebulosity is apparent. Malin's technique, however, brought out some faint underlying nebulosity which we could more clearly see in our CCD image (Figure 9.5).

When the enigmatic nature of this smudge galaxy was truly revealed, we christened it Malin 1, in honor of the technique that allowed its initial detection. As we shall see, Malin 1, in fact, is the largest spiral galaxy every seen, yet it is mostly invisible, because it is so diffuse!

Bagging the big one

The story of the discovery of Malin 1 is worth relating because it's a prime example of science as a discovery process and, indeed, accurately

Figure 9.5　　　　　The crouching giant, the barely visible disk of Malin 1.

reflects why some of us have become astronomers. In May of 1986, using the Palomar 200-inch telescope, Jeremy Mould and I took a spectrum of the nucleus of Malin 1. In most circumstances, galactic nuclei do not have emission lines but, in the case of Malin 1, we discovered an emission line spectrum. When a galaxy has an emission line, its redshift is fairly easy to determine as the line can be easily identified. But that was not the real surprise. The real surprise came in determining the redshift of this object. A rough reduction of the emission line spectrum at the telescope indicated recessional velocity of about 25,000 km/s. The recessional velocity of the Virgo cluster is about 1,200 km/s so this object was located at a distance well beyond the Virgo cluster which immediately made it interesting. The total angular size of the object on our CCD frame (see Figure 9.5) was approximately 2.5 arcminutes yet apparently it was 25 times farther away than Virgo. A quick scaling then indicates that if a galaxy like this was indeed in Virgo then its angular size would be a degree. If it were as close as the Andromeda Galaxy its angular size would be about 20 degrees and of course we would look right through it and never know it was there.

My colleague, Jeremy Mould, an optimistic Australian with incredible astrophysical instincts, thought we had a real winner here. I, not really believing in luck at stumbling across the largest spiral galaxy known (even though you can't really see it—Figure 9.5) was a good bit more skeptical. After all, this seemed like such an absurd structure for any spiral. I became convinced that we were seeing a background emission-line galaxy shining through an extended foreground dwarf galaxy in the Virgo cluster (statistical weirdness does happen in astronomy). In fact, I was so convinced that I bet a six-pack of beer with Jeremy Mould that this was the case. To win my bet, I would have to verify the existence of the foreground dwarf galaxy in the Virgo cluster. The best way to verify this would be to search for 21 cm emission from the neutral hydrogen associated with this foreground dwarf.

In September 1986 I left Caltech to take up a position at the University of Michigan in Ann Arbor. I was eager to resolve this issue with respect to Malin 1, so eager that in fact, right after I moved into my house in Ann Arbor, I got on an airplane and flew back to Arecibo in early October of 1986. I was determined to win my six-pack of beer. I tuned the receiver of the big dish to the Virgo cluster redshift and pointed at Malin 1 for days. That effort was futile. No signal. The object refused to yield and, if it was truly devoid of gas, then I could never prove my point, or, more importantly, win any beer.

Then, two days before the observing run ended, I had the astonishingly clever idea of actually tuning the observing frequency at Arecibo to the emission-line redshift that we had recorded at Palomar. Ten minutes later I got the surprise of my astronomical career. A very large signal at 21 cm was detected at the emission-line velocity. The signature of the 21 cm profile was exactly that of a rotating disk galaxy. I lost my bet, but in the process accidentally discovered Malin 1, an absolutely huge disk galaxy that was very, very diffuse. For the record, Malin 1 has a scale length that is 20 times larger than our Galaxy and and a central surface brightness of $I_0 = 26.0$. This is approximately 15 standard deviations fainter than Freeman's Law. And we had discovered the object accidentally! I knew then that the space density of LSBs had to be relatively large and that was the impetus to launch more systematic surveys for LSBs. Moreover, this was Disney's crouching giant, a direct verification of his long-ignored mathematical argument. We announced the discovery of this object in the summer of 1987 and with that opened up an entire new field of research in extragalactic astronomy.

The rest is history

Since that defining moment, I and my colleagues have continued to perform new surveys, using various techniques, to characterize the properties of this newly discovered class of galaxies. The importance of this newly discovered population cannot be overstated. The existence of LSB galaxies is a clear signal that the samples from which we select galaxies for detailed follow-up studies are incomplete, inadequate and biased. We have now discovered a few thousand of these previously missed diffuse galaxies that inhabit the same volume as those galaxies contained in the Zwicky catalog and which define the Hubble sequence. Intense study of the properties of these objects has produced several PhD theses which have built the foundation for the careers of other young astronomers. One of those former students, Stacy McGaugh (now on the faculty at the University of Maryland), has combined all the surveys made to date to produce a plot of the space density of galaxies (the number of galaxies per unit volume in the Universe) as a function of I_0. That result is shown in Figure 9.6 (color section).

These data are striking. Our survey results now indicate that up to 50% of the general galaxy population is in the form of disk galaxies with I_0 fainter than 22.0. Moreover, the space density remains flat out to the limits of the data. For galaxies with $I_0 \sim 25.0$ (the limits of our current data) the measured space density is millions of times higher than would have been anticipated based on Freeman's Law.

Thus, in just over a decade, a whole new population of galaxies has been discovered. These LSB galaxies are of cosmological significance and have properties which are different from those of their higher surface brightness counterparts which currently are dominant in extant galaxy catalogs. LSB galaxies offer a window into galaxy evolution which is different from that which has been traditionally used. Moreover, their mere existence and properties show the diverse array of evolutionary paths available to spiral galaxies. Still the quest for even more extreme forms of LSB galaxies continues. In the last two years we have used new wide-field CCD cameras to reach a detection limit of $I_0 \sim 27.0$ and have found a handful of new objects indicating there may be no real end to how diffuse galaxies can get. Such extremely diffuse galaxies are an extreme challenge to our understanding of how galaxies form and evolve, but their existence can no longer be ignored. In fact, a recent calculation by Impey and myself suggests that up to 80% of the entire baryonic content of the Universe may be contained in LSB galaxies. Clearly then, LSBs must be fully accounted for in any complete theory of galaxy evolution.

However, the discovery of LSB galaxies does very little to actually alleviate the dark matter problem. We already know that the stars, gas and dust contained in normal galaxies make up only 0.5% to 1% of the required mass to close the Universe (a closed universe is one that expands to some maximum radius and then contracts under the weight of its own gravity). Thus, while LSB galaxies are important locations for additional baryonic matter, unless they are made overwhelmingly of dark matter (and observations indicated they are not), then their presence does not substantially alter the fundamental result that material contained in galaxies, of any kind, is insufficient to close the Universe.

In the summer of 1998, the International Astronomical Union held Colloquium No. 171. The title was The Low Surface Brightness Universe. Approximately 100 astronomers from all over the world attended this meeting, which, fittingly, was held in Wales, the homeland of Mike Disney. At the meeting was a whole new generation of young astronomers who were working on various aspects of LSB objects. This was an indication that the field had now matured. Had this meeting been held in 1988, it could have been held in a small pub in that wilderness with Impey, Disney, and myself convincing each other that, in fact, we weren't really crazy and that invisible galaxies did exist in large numbers.

The discovery of LSBs has been gratifying, but to be honest it's somewhat ironic to be known as "the astronomer that studies invisible galaxies." The real triumph of this work lies in demonstrating how very little we may actually know about galaxies and the Universe, and that astronomy is a field that is based on discovery.

The Universe is far from revealing all of its secrets to us and we should never really be surprised at what we may discover. That's what this enterprise is all about—a voyage of human discovery that will take, perhaps, forever to complete. Along the way, new landmarks are found and characterized and new pages are written but ultimately its the voyage of discovery that drives us.

Suggested reading

Gregory Bothun, The Ghostliest Galaxies, *Scientific American*, Vol. 276, (1997), p. 40

C. Impey, Ghost Galaxies of the Cosmos, *Astronomy Magazine*, June 1996

J. Dalcanton, Ghost Galaxies: Outnumbering All the Rest, *Sky and Telescope*, Vol. 95, No. 4 (1998)